Social and Cultural Studies of Robots and AI

Series Editors
Kathleen Richardson
Faculty of Computing, Engineering and Media
De Montfort University
Leicester, UK

Cathrine Hasse
Danish School of Education
Aarhus University
Copenhagen, Denmark

Teresa Heffernan
Department of English
St. Mary's University
Halifax, NS, Canada

This is a groundbreaking series that investigates the ways in which the "robot revolution" is shifting our understanding of what it means to be human. With robots filling a variety of roles in society—from soldiers to loving companions—we can see that the second machine age is already here. This raises questions about the future of labor, war, our environment, and even human-to human-relationships.

More information about this series at
http://www.palgrave.com/gp/series/15887

Maciej Musiał

Enchanting Robots

Intimacy, Magic, and Technology

Maciej Musiał
Department of Philosophy
Adam Mickiewicz University
in Poznań
Poznań, Poland

ISSN 2523-8523   ISSN 2523-8531   (electronic)
Social and Cultural Studies of Robots and AI
ISBN 978-3-030-12578-3   ISBN 978-3-030-12579-0   (eBook)
https://doi.org/10.1007/978-3-030-12579-0

Library of Congress Control Number: 2019930716

Cover illustration: Besjunior/Alamy Stock Photo

This Palgrave Macmillan imprint is published by the registered company Springer Nature
Switzerland AG
The registered company address is: Gewerbestrasse 11, 6330 Cham, Switzerland

To my little daughter Izabela
for enchanting my world
with the magic of love

ACKNOWLEDGEMENTS

I am grateful to National Science Centre, Poland for funding the research presented in this book as a part of a grant no. 2015/19/D/HS1/00965.

I would like to thank Rachel Daniel and Madison Allums from Palgrave Macmillan for their kind assistance as well as to Murali Dharan Manivannan and Sneha Sivakumar for making the production process smooth.

I wish to express my gratitude to Kathleen Richardson, Cathrine Hasse and Teresa Heffernan, the Series Editors, for their comments and criticisms that have enabled me to make this book better. I am also deeply grateful to David J. Gunkel and Jennifer Robertson for their valuable feedback.

I would like to extend my thanks to my colleagues from the Department of Philosophy at Adam Mickiewicz University in Poznań. Karolina M. Cern, Mikołaj Domaradzki and Tomasz Rzepiński provided me with invaluable support at various stages of the process of preparing this book. Many thanks go also to Joanna Malinowska for reading and commenting on the draft version of the manuscript.

I remain indebted to Barbara Kotowa and Anna Pałubicka for their continuous support and advice.

I am profoundly grateful to my parents, Aleksandra and Henryk, for all their help.

I cannot possibly express my gratitude to my wife Paulina for always being there for me, keeping me sane and making me happy each and every day.

Finally, I would like to mention my beloved daughter Izabela, to whom this book is dedicated—for every single thing she does.

CONTENTS

LIST OF FIGURES

CHAPTER 1

Introduction

In most of our lives—or, hopefully, all of our lives—there are moments that can be described as "enchanting"—those magical times when someone or something puts a spell on us. The moment when you meet someone and fall in love at first sight. When you are contemplating a particularly beautiful landscape, breathing the fresh air. The time you devoured a particularly engaging book, immersed in its world and excited to find out what will happen next. Sooner or later, though, this initial enchantment often turns into disenchantment. A relationship becomes routine and the person we fell in love with seems to lose their charm. A big cloud of smoke from a nearby factory overshadows the landscape and pollutes the air. The book we have been reading with such excitement ends in an astonishingly disappointing way. The spell is broken, the magic is gone, the enchantment becomes disenchanted and we are disenchanted with it.

Luckily, there are also times when we experience re-enchantment: the old magic comes back or a new spell is cast. The people we love suddenly look and smile exactly the way they did when we met them for the first time. The wind blows the smoke away and a flock of birds rises into the sky from the forest as you gaze in awestruck wonder. The sequel explains the ending of the previous book in a way that renews your lost fascination.

© The Author(s) 2019
M. Musiał, *Enchanting Robots*,
Social and Cultural Studies of Robots and AI,
https://doi.org/10.1007/978-3-030-12579-0_1

Each of us has a story of his or her own enchantments, disenchantments, re-enchantments and the diverse relationships among them—personal experiences that define us as individuals. This book tells a story about such openings and closures and their interconnections in our contemporary culture, about the experiences that define us—at some level, to some degree and keeping great diversity in mind—as a contemporary "modern Western society". There are two main areas that I find crucially important in relation to the presence and influence of processes of enchantment, disenchantment, and re-enchantment. The first area refers to our attitudes and beliefs about humans and robots as well as to interactions and relationships between them—particularly the possibility of intimate relationships as well as actual interactions that are already taking place. The second area refers to modernity: the values, beliefs, and modes of thinking typical for what we label as "modern," our contemporary attitudes toward them, as well as our—probably still modern—accounts of premodernity and attempts to leave modernity behind and pursue in the direction of a kind of "beyond-modernity." The processes of enchantment, disenchantment and re-enchantment in both of these areas are strongly interconnected.

When we examine our past personal enchantments, disenchantments, and re-enchantments, we often come to the conclusion that they were not only—or maybe even not primarily—about the people or objects or any part of reality external to us, but were rather a matter of our own thinking. We notice that the routine in relationship was not the fault of the person we loved changing. Rather, it was a change in our own thinking that resulted in a different perception of that person. Or we notice that the landscape that enchanted us is not, in fact, as beautiful as what we see from our window every day, although we do not normally think of something that we see every day as capable of being enchanting at all.

A comparable situation occurs at the level of cultural enchantment, disenchantment, and re-enchantment—what and how we think, as well as why we think it, is a crucial matter. This book examines the question of thinking from a few different perspectives. First, I examine thinking as an illustration of processes of enchantment, disenchantment, or re-enchantment in the sense that I point to particular ways of thinking as an *examples* of each of the three abovementioned processes. Second, I view thinking as an expression of statements and attitudes that concern processes of enchantment, disenchantment or re-enchantment, particularly *diagnoses* that describe them, evaluate them and/or recommend their stimulation or implementation, or advise against it. Third, I look at thinking as a *source and condition* of enchantment, disenchantment and re-enchantment—the world or a part of it becomes enchanted, disenchanted or re-enchanted because we think about it in a particular way: in this book I focus on a magical thinking that makes enchantment and re-enchantment possible, in opposition to what we can call "modern rational thinking", which enables disenchantment—I emphasize that it is modern rationality, due to the fact that I do not think that only modern thinking can be recognized as rational, I do not consider magical thinking as simply irrational and I assume that magical thinking can be seen as rational in its own non-modern way. Fourth, I elaborate on thinking about thinking, that is, on fluctuations in *evaluations* of different modes of thinking, particularly magical thinking and its relation to modern rational thinking. I therefore discuss how our thinking contributes to enchantment, disenchantment, and re-enchantment (by being their expression), how it examines these processes (by diagnosing and evaluating them), how it constitutes these processes (magical thinking in the case of enchantment and re-enchantment; modern rational thinking in the case of disenchantment), and how it reflects on itself (that is, how we think about magical thinking and modern rational thinking).

This book is therefore about enchantment, disenchantment, and re-enchantment, about humans, robots and their interactions and intimate relationships, about magic, magical thinking and modern rational thinking, about modernity, premodernity and post-modernity, and finally about thinking: both the thinking about the abovementioned issues, and the thinking about the thinking itself, particularly about its diverse modes.

In Chapter 2, I discuss the question of *robots enchanting humans*. That is, the phenomenon of robots being perceived by humans as "magical" enough to develop intimate relationships with them. To examine this problem, I explore the debates in the field of robot ethics that concern intimacy robots, particularly sex robots and care robots. The chapter "Robots Enchanting Humans" thus discusses normative approaches that answer the question, "How should we think about robots, humans and intimate relationships?" I discuss both the arguments of those who are enthusiastic about intimacy robots, who look forward to and are optimistic about them, as well as the positions of skeptics, who consider intimate relationships with robots to be a serious danger. I do not attempt to decide which side is correct. Rather, I aim to understand what cultural beliefs and values—particularly these connected with ideas about humans, robots and intimate relationships—these approaches express. I therefore show how attitudes toward intimate relationships with robots are an expression of more general tendencies and transformations of our cultural ideas of humans and intimacy. To make this point, I explore sociological diagnoses of contemporary transformations of intimate relationships. Finally, I argue that if robots are being viewed as able to enchant humans and to become their intimate partners, it is because humans are becoming increasingly disenchanted, in a double sense. Humans and the relationships between them are becoming disenchanted, first, in the sense that they are perceived as possessing no unique, extraordinary and "magical" qualities that cannot be rationally calculated and engineered into the robots—this is a disenchantment *of* humans and relationships between them. And humans are becoming disenchanted in the second sense, that people and the intimate relationships among them are increasingly perceived as a source of disappointment—this is a disenchantment *with* humans and relationships between them. Humans are therefore increasingly seen as both non-unique and problematic, while robots are increasingly seen as possessing all the advantages of humans with none of their disadvantages.

In Chapter 3, I turn to the issue of *humans enchanting robots*, in the sense that humans tend to perceive and think about robots as something more than mere machines. Rather than look at prognoses about future human-robot intimate relationships and ethical, normative evaluations of them from the field of robot ethics to answer the question of "what we should think about robots, humans and intimate relationships," I explore two other fields of inquiry. First, the field of Human Robot Interactions (HRI), which describes actual human robots interactions that are already taking place to answer the question, "How do we think about robots (when we actually interact with them)?" Second, I discuss a specific part of robot ethics discussions that are aiming to rethink our thinking about humans, robots and relations between them, and in that sense ask the question: "How should we think about our thinking about robots?"—this rethinking is taking place partially due to the fact that the actual interactions and experiences of robots described in HRI do not correspond to our ideas of humans, robots, and intimacy expressed in the discussions examined in Chapter 2. Therefore, while the thinkers discussed in the chapter "Robots Enchanting Humans" mainly aim to adjust relationships with robots to our ideas of humans, robots, and relationships, philosophers debated in the chapter "Humans Enchanting Robots" attempt to adjust our thinking and ideas about robots, humans, and relations to our actual interactions with the machines.

What I find common in phenomena diagnosed by HRI studies and the philosophical propositions developed by contemporary influential thinkers is that both can be interpreted as expressions of the enchantment of robots, thinking about them as something more than mere machines. While HRI scholarship shows that humans *are* enchanting robots in actual, spontaneous interactions with them, some philosophers suggest that it might be valuable to take these experiences and interactions seriously and to reflect and understand these robots in a similar enchanting way. In other words, HRI studies show how humans are enchanting robots in *practice*, while some thinkers suggest that it might be productive to enchant them in *theory*, as well. I argue, moreover, that both of these enchanting approaches can be interpreted in terms of magic and magical thinking: I identify similarities and analogies between magic and magical thinking on the one hand, and HRI and philosophical ideas, on the other. In doing so, I draw on both classic and more recent anthropological, psychological, and philosophical accounts of magical thinking. I discuss reevaluations of the status of magic and magical thinking from the second part of nineteenth century to the present and focus on shifts in its relationship to modern rational thinking. Chapter 3 also addresses the question "How should we think about the magical and modern rational thinking?" I argue that the contemporary growth in the presence of magic and magical thinking might be a result of the disenchantment of modern rational thinking: this disenchantment includes a disenchantment *of* modern rational thinking (we no longer believe in its power) and a disenchantment *with* modern rational thinking (we are disappointed with the actual results of the domination of modern rational thinking).

Both chapters discussed above therefore discuss the "enchanting robots" signaled in the title. First one focuses on robots that are enchanting humans by taking on the status of participants in intimate relationships. Second focuses on how humans are enchanting robots, by thinking magically about them. The two chapters thus aim to show that robots are enchanting humans because humans are enchanting robots. In both cases, the enchantment is a result of a twofold disenchantment, of and with human beings and intimate relationships with them, as well as of and with modern rational thinking and its results, respectively. Therefore, Chapters 2 and 3 show how we tend to be less anthropocentric (by disenchanting humans) and less Western-centric (by disenchanting modern Western rationality). In some sense, these two chapters are two sides of the same coin, since they describe relations between humans and robots from two different perspectives: what robots do with humans (Chapter 2) and what humans do with robots (Chapter 3).

In Chapter 4, I discuss the disenchantment and re-enchantment *of,* *with* and *in* modernity. I examine accounts from diverse fields of social sciences and humanities—mainly anthropology, philosophy, religious studies, and sociology—which approach disenchantment and re-enchantment in at least one of the following ways: they diagnose the disenchantment or re-enchantment of modern Western world, they can be treated as an expression of such disenchanting or re-enchanting, or they recommend disenchanting or re-enchanting. I propose to understand disenchantment and re-enchantment as complementary processes that coexist and influence one another. In particular, I show how re-enchantment can be understood as a result of disenchantment. I want to avoid suggesting an either/or condition, in which the world is either disenchanted or re-enchanted. Rather, I claim that it is both disenchanting and re-enchanting. In the chapter "Disenchanting and Re-enchanting in Modernity", moreover, I show that the presence of magical thinking in reference to the robots discussed in Chapter 3 is a specific case of a broader issue of the presence of magic and magical thinking in contemporary Western culture. After showing the tendencies toward the disenchantment of humans (in the chapter "Robots Enchanting Humans") and the disenchantment of modern rational thinking (in the chapter "Humans Enchanting Robots"), which are the central values of modernity, I show that modernity itself is, in fact, undergoing a process of disenchantment, though, as a result, it is also undergoing re-enchantment.

In Chapter 5, in lieu of a conclusion, I briefly examine the limitations of the book as well as the limitations of most discussions about humans, robots and modernity. I will review the spectrum of questions the book addresses and explain why, instead of answering them, I try to understand the answers to them that have already been offered and pinpoint their origins, such as, for example, the cultural tendencies they express. I will also suggest that we might be putting too much confidence in our thinking and its impact. I therefore propose that important factors determining our thinking seem not to have been sufficiently examined. Finally, I point out to some future scenarios that might result from unintended consequences of our contemporary ideas, actions and tendencies we observe, specifically designing robots becoming eugenics and the meaning of life in the world of rampant automation.

To clarify and emphasize the statements made above as well as to shed additional light on the examinations made in particular chapters it is worth to explain the perspective from which this book is written, as well as the intentions and motivations that have driven its development. To understand this perspective, as well as its intentions and motivations it is therefore necessary to admit that the author of this book is a philosopher, particularly embedded in the field of the philosophy of culture. This specific aspect of the author's background determines the perspective and the method implemented in this book. Since philosophers do not have competences to conduct empirical research, this book does not present any such studies conducted by the author. What it does present are theoretical investigations and empirical studies conducted by others. The fact that the author represents philosophy of culture determines the way in which the investigations and studies have been selected and interpreted. Specifically, the author has attempted to choose the most influential, most discussed and most cited accounts of the problems described in this book, as he assumes that they do express, establish and/or sustain the beliefs and values that are typical of our contemporary culture, and by culture I understand a set of beliefs and values that is more or less commonly shared in a particular society. Therefore, the investigations and studies examined here are considered not simply as developments of individual scholars or research teams, but as a part of a general *Zeitgeist* that is present not only at the level of academia, but also outside of it. So, I believe that by discussing academic studies I do not only give a sense of the scholarly field, but also a sense of our culture. However, I do limit my considerations to academic papers and do not refer to sources such as press articles or science-fiction books, series and movies, since I believe that this former source offers the most clear and detailed image of the discussed problems as a part of our culture. And this in fact motivates my intention to mainly describe and compare diverse positions without evaluating them, which is particularly visible in Chapter 2 where I examine enthusiasts and opponents of sex robots and care robots. I am certain that most persons who have some strong beliefs on this issue would claim that in this chapter the position with which he or she agrees is not sufficiently investigated and not discussed deeply enough, or even marginalized. However, I also believe that after a closer examination such persons would admit that their view is presented in an analogous way as opposite positions and therefore that these positions are presented in a balanced way, and moreover that such persons would notice that their views are discussed not only as a voice in a particular discussion, but also as a part of a more general cultural tendency. Therefore, the aim here is rather to understand, than to evaluate. Instead of deciding who is right I want to decide who is who: where does he or she come from, what cultural values motivates him or her, what tendencies do they express, establish or sustain and where it may lead to. As a philosopher I try to generalize and make the discussion more abstract, but as a philosopher of culture I do not go to the heights and depths of metaphysics and objective realities but prefer to stay at the level of culture and inter-subjective beliefs and values. Those who are strongly engaged in these discussions might feel disappointed with such a neutral approach, but I believe that it offers something for them as well. Specifically, I think that understanding the position that is opposite to our own in a better way enables us to criticize it more accurately and severely on the one hand, and to understand better our own position on the other.

One could argue that the approach described above does not offer anything new since I only summarize investigations and studies conducted by others without even evaluating them. Below I will emphasize the aspects of the book that I consider as a kind of a novelty, by pointing out what discussions and literature the particular chapters of the book respond to and what do they add to them. Chapter 2 responds to all literature on intimacy robots (e.g., sex robots and care robots). The main novelty of this chapter is that it connects the topic of sex robots and care robots together and shows similarities between discussions about them (other approaches examine only one of these types of robots), discusses them from a relatively neutral point of view and considers the arguments of both sides as expressions of broader cultural tendencies (other approaches are either for or against), and links these matters to sociological studies about contemporary transformations in intimate relationships between human beings.

Chapter 3 responds to Human Robot Interaction studies and philosophical reflections about rethinking our thinking about robots on the one hand, and both classical and contemporary accounts of magical thinking and magic developed by anthropologists, psychologists and philosophers on the other. The novelty is that it applies magic and magical thinking to the area of interactions with robots and offers an explanation of some results produced in HRI, in addition to pointing out the presence of some aspects of magic in some philosophical investigations about robots. While there are other works that connect magic and technology, to the best of my knowledge none of them refers to particular cases studied in HRI nor point to the fact that some aspects of philosophical investigations on robots involve magical thinking as well.

Chapter 4 responds to literature from the diverse fields of the humanities and social sciences that offers some account of modernity, particularly those that discuss this issue in terms of disenchantment and/re-enchantment. The relative novelty here is that it emphasizes the complementarity of disenchantment and re-enchantment of the world, instead of treating them as mutually exclusive. Moreover, this chapter embeds the problem of robots and our magical attitudes towards them in a more general discussion on disenchantment and re-enchantment in modernity—putting robots in this context seems rather uncommon and therefore might be considered as relatively new. Finally, two scenarios discussed in Chapter 5 are referring to problems that are very scarcely present in the discussions about robots and I offer views opposite to, or at least different than these that have been developed before.

To summarize, I believe that the distinctive character of the book comes from pointing out significant links between these three rather rarely connected areas. To put it briefly: I show that we seriously contemplate intimate relationships with robots because we think magically about them, and that the presence of magic and magical thinking in contemporary culture requires us to rethink our understanding of modernity and of the disenchantment of the world.

CHAPTER 2

Robots Enchanting Humans

In this chapter, I discuss the topic of robots enchanting humans. That is, robots that are perceived by humans as magical enough to develop intimate relationships with them. I argue that robots are enchanting humans because humans are becoming increasingly disenchanted—in the double sense. First, humans and the relationships among them are becoming disenchanted in the sense that they are perceived as possessing no unique, extraordinary, and "magical" qualities that cannot be rationally calculated and engineered in robots—this is the disenchantment *of* humans and the relationships among them. Second, humans are becoming disenchanted in the sense that people and their intimate relationships are increasingly perceived as a source of disappointment—this is disenchantment *with* humans and the relationships among them. Humans are therefore increasingly seen as non-unique and problematic, while robots are increasingly seen as possessing all the advantages of humans and none of their disadvantages.

To examine the abovementioned issues, I refer to the field of robot ethics, particularly to the discussions that are answering the question of how we should think about lovebots, sexbots, and carebots, as well as the question of how we should think about humans and their intimate relationships. I examine both positions that are enthusiastic about the intimate relationships with robots and those that are skeptical about them. Nevertheless, my aim is not to decide which side of the discussion is right, but to understand where these positions come from, and where they might lead, by embedding them into more general cultural

© The Author(s) 2019
M. Musiał, *Enchanting Robots*,
Social and Cultural Studies of Robots and AI,
https://doi.org/10.1007/978-3-030-12579-0_2

transformations of intimacy studied by sociologists, as well as by pointing out some of the general axiological, epistemological, methodological, and ontological issues that these discussions involve.

2.1   Preliminary Issues

Before I get to the bottom line, I must pause to briefly examine some introductory issues. First, I define intimacy and intimate relationships; second, I discuss the status and types of the robots that are an object of interest in this chapter; third, in order to justify this topic, I address skepticism concerning the possibility of intimate relationships with robots; and fourth, I describe the area of discussion that interests me and introduce the order in which I examine it.

To define intimacy, I will begin by referring to previous attempts to provide a definition. First, I would like to point to Ronald Arkin's and Jason Borenstein's approach, in which they characterize intimate relationships by saying that their definitional feature is the fact that they involve love, and that "love, in a broad and encompassing sense of the term, is an essential component of intimate relationship" (Borenstein and Arkin 2016). Similarly to Arkin and Borenstein, I will not try to define what love is, since the topic is extremely broad and discussing it requires a separate book. I will simply make reference to the common-sense understanding of love as a deep emotional bond, and consider pointing to its presence in intimate relationships as a first step of explaining what those intimate relationships are. To take the second step, I need to refer to investigations that do not concern robots, but are devoted to issues of intimate relationships in general.

In a book about intimate relationships in modern societies, Lynn Jamieson distinguishes four types: couple relationships, sexual relationships, family relationships, and friendships (Jamieson 1998). Jamieson's typology could probably be criticized for many reasons (for example, for its ethnocentric character: in many non-Western societies we can observe polygamous intimate relationships, which—obviously—are not couple relationships). Nevertheless, I do not want to get into such detailed matters. I only need to point out that intimate relationships can be distinguished—in a rather common-sense fashion—into couple ("romantic"), sexual, family, and friendship relationships. This seems to be compatible with Arkin's and Borenstein's approach, since all four types of intimate relationships seem to involve (some type of) love.

To proceed further in defining intimate relationships, I would like to point to the fact that some scholars interested in intimacy emphasize that intimate relationships involve not only love, but also care (Hochschild 1983, 2003, 2012; Hochschild and Ehrenreich 2013). Without interrogating the relationship between love and care, I would like to claim that intimate relationships can involve both.

Intimate relationships between people and robots are thus, by definition, also relationships that typically involve love and/or care and can be differentiated, following Jamieson, into couple relationships, sexual relationships, family relationships, and friendships. The social (or sociable) robots that constitute the topic of interest in this chapter are above all love and sex robots (which are designed to participate in couple or "romantic" and sexual relationships), care robots (which are designed to take care of elderly people or children in the place of family members or human nurses or nannies) and other kinds of companion robots (artificial companions), which are designed to take part in family or friendly relationships. To put it another way, this chapter is interested in the question of robots being used in relationships that are most commonly understood as being "essentially human," "deep," and require a "human touch"—something that is typical only for humans, makes human interactions exceptional and brings a kind of magic into them. In other words, as I have already pointed out, I would like to talk about how robots may be able to enchant human beings to the point that we can form intimate relationships with them.

It is no surprise that in contemporary Western culture the idea of using robots for such purposes is seen as controversial and is debated, not only within academic discourse, but also in a wide range of media, from newspapers to television and movies. Nor is it unusual for these two discourses to mix, in media reports about Kathleen Richardson's "Campaign Against Sex Robots," for example, or about the fact that David Levy's lecture in Malaysia about love and sex with robots was cancelled under pressure from the Malaysian police. Nevertheless, in what follows, I draw primarily from the academic discourse, since I believe that it represents most clearly the cultural tendencies that concern the attitudes towards intimacy robots and intimacy in general.

There is another thing I need to make clear about this chapter: the robots I describe here are neither persons nor subjects in either the ethical or the ontological sense—in other words, these intimacy robots are neither conscious nor feel pain, and do not possess any other objective qualities that—according to most (though not all) scholars—are the minimum criteria for those ascribed the status of person or subject. When, in this chapter, I discuss the question of whether intimacy robots are, or could be, or should be subjects or persons, on the one hand, or objects/ tools, on the other, I therefore mean (like all other scholars engaged in this aspect of the discussion) to ask this question specifically *from the subjective point of view of the people who interact with them*, not from an "objective" point of view focused on "real" properties. In the following chapter ("Humans Enchanting Robots") I problematize this issue from different angle by referring to Coeckelbergh's and Gunkel's philosophy, and in the last chapter of the book I briefly discuss the possibility of the robots who are subjects and persons in every possible way.

Before I examine the main concerns in the debate, I want to deal with a possible disbelief that may appear when intimacy robots are discussed. Many people, quite frankly, may doubt that such problems are even worth investigating, since they doubt that intimacy robots will be implemented in practice. Other skeptics might claim that robots will never be sophisticated enough to attract people, while other disbelievers might believe that no matter how sophisticated robots may become, human beings will not be attracted to them, either because of their artificiality, non-humanness, and object/tool status or because such intimacy will be restricted by law or customs and morality. The following three points should serve to weaken such doubts.

First, intimacy robots are no longer the stuff of science fiction. They already exist and are in use. The Japanese and South Korean governments plan to solve their population crises by using carebots rather than using immigrant labor, as many other countries have. It seems that some European countries with similar problems may follow their lead (Borenstein and Pearson 2010; Sharkey and Sharkey 2012). In fact, robots like Paro, a seal-like artificial companion designed as a therapeutic instrument for the elderly (for detailed information about Paro, see Pfadenhauer and Dukat 2015), are already in use in many nursing homes for the elderly throughout Europe. Moreover, some of the existing prototypes of humanoid robots, androids, and geminoids—especially those built by Hiroshi Ishiguro and David Hanson—closely resemble humans in their appearance, movements, and even their communication skills. Robots designed for sex also already exist (although they are far less sophisticated than Ishiguro's and Hanson's robots). One example is Douglas Hines and his company—significantly named "True

Companion"—who developed Roxxxy, often called the first sex robot. Skepticism about the impossibility of building robots sophisticated enough to take part in intimate relationships is thus invalid: such robots already exist and are constantly being improved.

Second, it seems that demand for intimacy robots is quite robust. This can be seen in the results of polls and surveys performed in Western countries (Enz et al. 2011; Fong et al. 2003; Scheutz and Arnold 2016; YouGov 2013). They show that a significant number of people are interested in purchasing lovebots, sexbots, carebots, or other artificial companions, which means that a market niche exists, and it is reasonable to suppose that it will be filled with an appropriate supply—simply because, as Borenstein and Arkin claim, it seems to be a profitable business (Borenstein and Arkin 2016). This might be why some surveys show that people in technical professions are more positive and enthusiastic about the implementation of social and intimacy robots (Enz et al. 2011). In the next step, I show that the enthusiasm for intimacy robots evident in surveys can also be seen as a declarative expression of practical attitudes that can be already observed.

A third point is that it would be surprising if people would *not* animate, anthropomorphize, and emotionally attach to robots and treat them as intimate partners, since they have done that with far less sophisticated inanimate objects for centuries. The ancient myth of Pigmalion is a great illustration of this fact, showing that not only East Asian (Laue 2017; Levy 2007b; Yeoman and Mars 2012), but Western culture (Burr-Miller and Aoki 2013; Holt 2007; Musiał 2018; Sullins 2012), as well, is open to this tendency. Brothels currently exist, both in Asia (Culbertson 2017) and Europe (Nevett 2018), where sex dolls have replaced human beings. It is important to notice that in the cases mentioned above we can observe both animation or anthropomorphization, on the one hand, *and* emotional attachment, on the other. It is worth distinguishing these cases from animation and anthropomorphization *without* emotional attachment (which, as Fritz Heider and Marianne Simmel showed in the first half of the twentieth century, may refer to even simple geometrical objects (Heider and Simmel 1944), and from emotional attachment without animation and anthropomorphization (which, in fact, pertains to most people, to the extent that they possess any objects they consider particularly precious). Technology has therefore only strengthened this tendency. People tend to animate, anthropomorphize, and emotionally attach not only to robots—even unsophisticated ones (I discuss this in detail in the next chapter). They engage likewise with chatbots such as ELIZA and, indeed, with computers in general (Borenstein and Arkin 2016; Reeves and Nass 1996; Sparrow 2002; Yeoman and Mars 2012). Whether animation, anthropomorphization, and emotional attachment to technology differ from the treatment of other inanimate objects in quantity, in quality, or not at all, is a topic for a separate discussion.

Simply put, if it is natural (in some sense) that people animate, anthropomorphize, and emotionally attach to inanimate objects, especially to technological devices, it does not seem convincing to suggest that they will not do so with sophisticated robots designed specifically to stimulate such behavior (Sharkey and Sharkey 2006).

There are thus serious reasons to believe that intimate relationships between robots and humans are a highly possible scenario. Nevertheless, I would like to emphasize that even if human beings never actually become enchanted by robots to the extent of forming intimate relationships with them, I still consider discussions that seriously take such scenarios into account to be important and telling since they can tell us a lot about our ideas about robots, humans, and intimate relationships in general. Put simply, by examining this scenario and our attitudes toward it we can learn about ourselves. Actually, the very fact that many of us think this scenario is possible seems to constitute significant information about our culture.

As I mentioned above, I will focus on lovebots and sexbots, on the one hand, and carebots, on the other. What I find surprising in the discussions about those kinds of robots is that they engender two separate discourses. Scholars who talk about lovebots and sexbots very rarely raise the question of carebots and those who are interested in carebots typically remain silent about lovebots and sexbots. I aim to connect and combine those two discussions, not only because they both involve intimate relationships, but also, mainly, because they involve very similar issues and disagreements and complement each other.

It is important to notice that issues connected with intimacy robots can be discussed, as Matthias Scheutz emphasizes, from various "axes of evaluation," including "single individual vs. partnership or relationship, personal effects vs. large-scale social dynamics, judging by effects vs. fundamental dignity or norms about sex and relationships" (Scheutz and Arnold 2016). Similar distinctions are also proposed by Robert Sparrow (2002; "effects on the user" and "alterations of human-human relationships") and John Danaher et al. (2017; individual good and social good). I would therefore first like to examine the issues and discussions connected with individual good, user perspective, and personal effects, and afterwards examine problems and debates linked to the social good, societal perspective, and effects on society (especially on the quality of intimate relationships). There are also problems—such as the question of "human touch"—that will pervade throughout the whole chapter on both levels of discussion. This is obviously a result of the fact that both areas are intertwined. I still believe, however, that it is useful to distinguish them.

In both areas of discussion, there are arguments "for" and "against" intimacy robots (as well as counter-arguments) delivered by groups I will call *enthusiasts* and *skeptics*. It is worth pointing out at the very beginning that, regarding issues at the level of individual good, enthusiasts are on the offensive (they provide arguments "for" intimacy robots) while skeptics remain on the defensive (developing counter-arguments to enthusiasts' claims). At the level of social good, however, skeptics are on the attack (they present arguments "against" intimacy robots) while enthusiasts mostly defend (by forming counter-arguments). This is one reason why differentiating between these two levels is productive and reasonable. Nevertheless, I need to emphasize, once again, that for each problem I try to give justice to both enthusiasts and skeptics and I do not aim to adjudicate which side is right. When I criticize any of the positions I examine, it is only because it is criticized by opponents (I only reconstruct someone else's critique) or because I see some general errors that make the particular position unclear, inconsistent, or getting in the way of productive discussion. Instead of deciding who is right in his or her enthusiasm or skepticism, I rather try to understand where this enthusiasm or pessimism comes from. I try to understand what contemporary Western culture can tell us about the debate on intimacy robots, and what the debate on intimacy robots can tell us about contemporary Western culture. Rather than choose sides, I try to understand both, and to identify the main—ethical, axiological, methodological, and ontological—disagreements and problems that are (implicitly or explicitly) expressed in the discussions, but that are also valid and important in contemporary Western culture, in general—especially the problem of enchantment, disenchantment, and "human touch."

Before getting to the central point, I would like to briefly mention a few issues that are not important in my examination, but remain crucial for the general discussion of intimacy robots. First, I will not wade into the question of whether intimate (social) robots are something utterly new and whether the way people do and will interact with them is unprecedented (de Graaf 2016; Scheutz and Arnold 2016) or, on the other hand, whether there is nothing radically new about them and they are simply more sophisticated tools made for human use (Levy 2007a). Second, since I do not want to take a position on what should be done about intimacy robots, I do not engage in the discussion about how to design them, how to decide how to design them (de Sio and Wynsberghe 2016; Wynsberghe 2013), and how to practically evaluate their functioning effects (Stahl and Coeckelbergh 2016). Third, I will not attempt to categorize intimacy robots, as is often done with care-bots (Bendel 2015; Sharkey and Sharkey 2012). Fourth, I am not interested in historical development of robots and automations (Devlin 2018; Truitt 2015).

2.2   Individual Good, User Perspective, Personal Effects

In discussions of the impact of intimacy robots (particularly lovebots, sexbots, and carebots) on their individual good from the user perspective, with the focus on personal effects, three main areas of investigation seem to be crucial. First, issues of security, safety and privacy; second, issues of autonomy and freedom; third, an issue of positive experiences.

2.2.1   Security

Issues of security have multiple levels and all the levels are intertwined and mixed with each other. I start with the problems that are least important for my purposes (although they are extremely important in general) and then I get to those, which are crucial part of my argument. I briefly discuss issues of physical safety, then look at privacy and data security; finally, I examine the issue of ontological security.

Physical safety is a crucial element in discussions of robots—not only intimate ones, but also other kinds (self-driving cars, for example). In the case of intimacy robots, which, by definition, remain in close contact with their users, this question is especially important, particularly in the case of robots meant to care for the elderly or for children. One issue obviously concerns avoiding any malfunction that might threaten the users of such robots. While skeptics claim that it is not possible to avoid malfunctions and therefore robots should not be implemented in areas where they might pose a threat for humans, enthusiasts believe that it would be satisfactory enough if robots pose a statistically lower threat than do human beings themselves (for example, if self-driving cars would cause fewer accidents than human drivers). This issue, however, remains a technical problem and I will not discuss it further here. A second issue concerns the balance between the ability of robots to guarantee users' physical safety, on the one hand, and the possibility that robots may limit the freedom and autonomy of the user, on the other.

Amanda and Noel Sharkey pose some illustrative questions in regard to this: if an elderly person wants to drink alcohol, for example, something that is not good for his or her health, how strongly could or should a robot act to prohibit this choice? If an elderly person wants to walk onto the road into heavy traffic should robot stop this person (and provide safety) or not (and provide autonomy)? (Sharkey and Sharkey 2011). This extreme example shows the importance of the balance between physical safety and personal freedom and autonomy. This problem is not entirely new, however: it is already present in the context of human caretakers, to a similar degree. The difference, of course, lies in the fact that human caretakers can negotiate the balance with those who are cared for, while robots need to be programmed to perform specific actions chosen from some binary repertoire without the ability to create in-between options.

Another dimension of the physical safety question mainly concerns sex robots. Some sexbot enthusiasts argue that if these devices replace prostitutes and casual sexual partners, they will increase their users' physical safety by lowering the number of cases of sexually transmitted diseases (Levy 2007b; Yeoman and Mars 2012).

Privacy and data security are other important issues at play with robots in general and intimacy robots in particular. One of the main problems here are the personal feelings of the users, especially elderly people using carebots. Assuming that carebots would help the elderly bathe or change their clothes, would their users feel that their privacy is being invaded? Would this sense of invasion be higher or lower than when those tasks are performed by a family member or hired human caregiver? It seems that this problem is, in fact, a matter of individual preferences and cannot be solved through any general decision. The second problem that concerns data security seems to exceed the area of personal preferences and demands. One of the functions of carebots will be recording the behavior of the elderly or of children, so that the primary human caregivers can remotely control the situation or simply check on what the child, parent, or grandparent is doing in their absence. The obvious question is, what should happen with these recordings? Who has right to watch them and how long should be they preserved? How can this data be protected from theft? Such questions demand specific regulations in the law in order to protect the right to privacy of robot users, especially the most vulnerable (children and the elderly) (Calo 2011).

As I have said, I will not get into further details about issues of physical safety, privacy, and data security. An enormous number of experts have examinations in that matter, and I have nothing new to add (Borenstein and Pearson 2010; Coeckelbergh 2010; Sharkey and Sharkey 2011, 2012). I see these issues as essentially practical concerns, while my purposes here are mainly theoretical. I would therefore like to shift my attention now to the issue of ontological security.

The concept of ontological security was introduced by Anthony Giddens (1991, 1992). According to Giddens, ontological security is connected with a subjective feeling of having a stable, meaningful life. This sense is maintained by comfortable routines, and can be eroded by chaos, anxiety and risk. Giddens claims that ontological security is particularly important and precious in the late modern era, since constant change and lack of stability are one of the main features of late modernity in general, and of late-modern intimate relationships in particular. The British sociologist labels contemporary intimate relationships "pure relationships" and claims that they are freed from cultural restrictions and societal pressures and depend on free choices of individuals, and, at the same time, are by definition temporary contracts. Similar diagnoses—on both the general and the intimate levels—are offered by Urlich Beck and Elisabeth Beck-Gernsheim (Beck 1992; Beck and Beck-Gernsheim 1995) as well as by Zygmunt Bauman (2000) (to my more detailed examination of these and others accounts of contemporary transformations of intimacy, see [Musiał 2013] in English, or [Musiał 2015] in Polish). Simply put, contemporary transformations of intimacy provide people with more freedom to shape their relationships, but also with more uncertainty and less stability.

The development of intimacy robots seems to be a response to this state of affairs. To put it bluntly, intimacy robots do not cheat, do not leave, do not disappoint (Levy 2007a, pp. 132–139)—they are therefore capable of providing ontological security. What is more, intimate relationships with robots eliminate the risk of unwanted pregnancy, which partially refers to physical safety but also to ontological security since the risk of pregnancy is often experienced as a source of anxiety. Moreover, intimacy robots can be a therapeutic for people whose ontological security has been broken by unsuccessful human relationships. David Levy suggests that robots will be more efficient than Prozac in that regard (Levy 2007a, p. 105). In fact, if I may draw on some anecdotal evidence, stability and lack of risk is one of the main advantages emphasized by people who prefer to live with dolls than with human beings. One of them claims that not many of his childhood friends are his friends in the adult life, while dolls will be with him forever (Holt 2007). Arkin and Borenstein wonder whether such security will not become boring (Borenstein and Arkin 2016). It is rather easy, however, to answer this doubt from the perspective of an enthusiast. First, robots can be programmed in various ways that guard against routine and boredom (the sexbot Roxxxy, for instance, has multiple "personalities," ranging from Frigid Farrah to Wild Wendy). Second, there is no need to stick to one robot; one can always buy a new model that provides new experiences, without any of the guilt or remorse (not to mention other difficulties) that naturally accompanies divorces and breakups with human beings. The matter of ontological security is significant also in the case of carebots: empirical studies show that contact with intimacy robots reduces stress, anxiety, and loneliness in the elderly (Sharkey and Wood 2014). It therefore seems that robots can be seen—at least by their enthusiasts—as a cure for a contemporary disease of lack of ontological security. As Sherry Turkle puts it, technological devices like robots deliver a promise that we will never be left alone, that there is always someone/something by our side or, at least, someone we can immediately get in touch with (Turkle 2012). Robots and technology are not the only way to achieve this promise, but they do seem quite promising.

2.2.2   Autonomy and Freedom

Without getting into philosophical discussions about the nature of freedom and autonomy, I would like to define these terms, with Sorrell and Draper (2014, p. 192), as simply "being able to set goals in life and choose means" and complement them with notions of "independence—being able to implement one's goals without the permission, assistance or material resources of others" and "enablement—having or having access to means of realizing goals and choices." I will now examine the debate between enthusiasts who claim that intimacy robots support the fulfillment of the abovementioned values for their users and skeptics who believe that intimacy robots may lead to an erosion of those values.

In the case of lovebots and sexbots, their enthusiasts and proponents emphasize that, unlike human partners, they do not limit freedom and autonomy at all, since they can be tailored to users' needs (Levy 2007a, pp. 132–139, 157). Lovebots and sexbots can be programmed to (simulate) love toward their users, without requiring any love in return. To put it bluntly, robot will not tell us what we can/should or cannot/should not do—it does not limit our range of options. Robots are not envious, do not feel bad about being cheated on, have no preferences about what to watch on TV, have nothing against drinking beer or going shopping. Again, I would like to bring in anecdotal evidence from doll owners who explicitly claim that dolls are better than people since they do not limit a person's freedom in any way—one doll user claims that he feels like his own god living in his own world (Holt 2007).

The same expansion of freedom and autonomy (as well as independence and enablement) also seems to occur when elderly people use carebots. Elderly people, who may be impaired because of their age and/or because of various diseases have quite limited autonomy and freedom and are heavily dependent on other human beings. Carebots, especially assistive robots, would increase their autonomy in the sense that they would be empowered both physically (changing clothes, taking a bath) and mentally (reminded to take pills) and be enabled to deal with some tasks without the help of human caregivers (Decker 2012; Sharkey and Sharkey 2011). In that sense, robots can function as a "physical and cognitive prosthesis" to "liberate" the elderly "from some of the burdens of dependence" (Borenstein and Pearson 2010, p. 282). Finally, the increase in autonomy, freedom, independence, and enablement is connected with the fact that some carebots (e.g., artificial companions, such as Paro), rather than *providing care*, provide companionship and actually *require care* themselves. The feeling of being a caregiver, of being in control of someone or something seems to stimulate the fulfillment of the abovementioned values (Misselhorn et al. 2013).

Yet another skeptical point of view toward the impact of intimacy robots on autonomy, freedom, and independence is possible, and it is most starkly represented by Matthias Scheutz (2011). Scheutz believes that most people will not be able to treat robots as simple tools and will develop strong emotional attachments toward them—this belief is founded on observations of people's already-existing emotional attachment to relatively simple robots such as Roomba, AIBO, or Packbot (see Chapter 3 of this book for further details). Scheutz's main worry is that people will become psychologically dependent on their robots. The negative consequences include getting addicted to robots. On the one hand, this worry seems legitimate, since addictions are generally negative and we already know that addictions to technology (TV, video games, etc.) are a highly possible scenario. On the other hand, we may ask whether strong attachment to someone or something that is one's object of affection should be labeled "addiction" at all. Another illustration provided by Scheutz refers to the influence robots may have on their users. In the case of intimacy robots, which are potentially objects of strong emotional attachment, including love, we may suppose that those robots—similarly to human partners—will be able to persuade their users to do things they wouldn't do otherwise, and, eventually, to manipulate them. Since, in this chapter, I am specifically discussing robots that do not have consciousness and intentionality, such a concern may appear unwarranted. Scheutz, however, emphasizes that robots would not manipulate users by themselves, but may become a means by which their producers and their potential partners would perform the manipulation—for example by programming a robot to praise some specific commodity and to persuade the user to buy it. Such attempts would be far more effective than TV commercials or billboards, since the robot's claims would not be seen by the user as a boring and/or aggressive commercial, but rather as the precious opinion of a loved one. An already-existing example of this risk can be found in China, where tens of thousands of (mainly) men have been lured into spending huge amounts of money by chatbots they thought were real women (Allen 2018). This issue is obviously also connected with the problem of deception discussed in the section of this chapter entitled "Positive experiences."

The impact of intimacy robots on the autonomy and freedom of their users (as well as on independence and enablement) thus seems potentially to be a double-edged sword. On the one hand, lovebots and sexbots can liberate people from the burden of dealing with the subjectivity of other human beings, especially their beliefs and desires, and carebots may empower the elderly and make them more independent and autonomous. On the other hand, there is a risk that people will attach to robots too strongly and in effect will become psychologically dependent on them as well as vulnerable to various forms of manipulation. It seems that one of the main factors that may determine whether enthusiastic or skeptical scenarios is more possible is the way in which people will perceive and treat robots—whether they will treat them as subjects and persons and develop strong emotional attachments, or rather as objects and tools that are simply used in order to fulfill one's needs (the abovementioned dilemma therefore does not refer to the objective status and features of robots, but rather solely to the subjective attitudes of the users of robots). I will elaborate on this issue in the remainder of this chapter as well as in the next chapter.

2.2.3   Positive Experiences

Now I would like to examine how those who are enthusiastic and skeptical about intimacy robots evaluate the positive experiences, such as satisfaction, joy and pleasure, derived from interactions with robots. To begin with, it can be said that while enthusiasts emphasize the fact that this involve providing users with real and actual positive experiences such as satisfaction, joy and pleasure, the skeptics point to the fact that these experiences are in fact a result of deception, a self-illusion, and a self-fraud.

A brief and clear way to present the bottom line of the enthusiasts' arguments is to look at a statement by David Levy, who believes that lovebots and sexbots (will) provide "great sex on tap for everyone 24/7" (Levy 2007a, p. 310). In other words, intimacy robots are unlimited, easily accessible source of positive experiences associated with intimate relationships. Proponents of this view emphasize that intimacy robots can not only be as good as human beings at delivering intimate pleasure, but can be even better than them. Robots can be programmed, for example, to possess an assemblage of sexual techniques beyond the capacity of most human beings due to its broad or extreme character (Levy 2007a, p. 22; MacKenzie 2014), and can simply offer a "combination of social, emotional, and intellectual skills that far exceeds the characteristics likely to be found in a human friend" (Levy 2007a, p. 104). Moreover, robots offer a great variety of intimate experiences, particularly because the "personality" of particular robots can be changed, or the particular robot can be replaced by another one (Levy 2007b). It is also important that this stream of pleasure comes without any strings attached—it is constantly available, there are no conditions one has to meet to access it (apart from the ability to purchase the robot) and—as I have already mentioned in the section about security—there are none of the risks commonly associated with intimate relationships with human beings, such as the risk of pregnancy or STDs, not to mention performance anxiety or the risk of being rejected or abandoned (Yeoman and Mars 2012). It is also important to point out that—according to enthusiasts—with intimacy robots, access to intimate pleasures becomes more egalitarian, more accessible to those, who for various reasons (shyness, unattractive physical appearance) are not able to find human partners for intimacy (Danaher et al. 2017; McArthur 2014).

As for the carebots, it also needs to be emphasized that most of the positive effects they may have on their users (the recipients of care)—including the reduction of stress, anxiety, and feelings of loneliness—seem to be connected with the fact that robots enchant them in the sense explained at the very beginning of this chapter. That is, the robots take on the appearance of something more than mere machines; they become friends, companions, and caregivers. Simply put, their potential to provide a significant number of positive experiences (and/or a reduction of negative experiences) is possible because the users are enchanted by their robots—just as the users of lovebots and sexbots are.

Skeptics, however, make at least four arguments against optimistic and enthusiastic attitudes toward the character of the positive experiences and enchantment that they are consequences of. One argument is that such enchantment and positive experiences derived from it are only temporary. A second argument calls attention to the limits of enchantment, to what positive experiences robots *cannot* provide (e.g., the fulfillment of a second-order desire-to-be-desired). A third argument concerns deception. A fourth involves the fact that the enchantment cannot provide users with actual experience of "human touch"—I will elaborate on this issue in the last section of this chapter.

The first argument is based on the belief that the enchantment of users by robots and the positive experiences derived from it are only a temporary novelty effect, which will not last long. Robots, skeptics contend, are something more than machines and provide new experiences and other positive effects only at the beginning. Eventually, they became just another boring commodity, a tool used for particular purposes, like any other (Sparrow and Sparrow 2006). This belief is supported by anecdotal evidence, such as a situation from a facility for the elderly in Kyoto, where residents were excited about their robot companions only for a month and became utterly uninterested in them afterwards (Sharkey and Sharkey 2010a).

This argument thus claims that not only enchantment, but also the positive experiences emphasized by enthusiasts, are of limited duration. At least two counter-arguments can be made in response. One is that the skeptical argument may take into consideration only currently-existing, unsophisticated robots, whereas the more highly developed lovebots, sexbots, carebots and artificial companions of the future will likely be able to sustain the enchantment. A second counter-argument is that even if the enchantment provided by one particular robot is only temporary, this is similar to the enchantment provided by anything else, including human beings—especially intimate partners. But robots offer the opportunity of replacing one model with another without any of the complications typical of divorces or any breakup between human beings. Nevertheless, it does seem that a proper examination of the skeptic's "novelty effect" argument demands time and detailed empirical study.

Dylan Evans (2010) introduces the second argument mentioned above most clearly. He claims that even if robots can fulfill our first-order desires, they cannot fulfill our second-order desires, particularly the desire to be desired, the experience of someone freely and without any coercion wanting us. Evans notices and emphasizes the paradoxical character of this desire: people want someone who is always free to leave, but never does. Robots, according to Evans, do not meet the first condition: they do not have the free will and intentionality that enables them to choose to be with someone or to leave someone. Of course, enthusiasts could counter that robots do not have to *really* have free will and intentionality, only to simulate having them, behaving *as if* they have them. This inevitably leads to a third skeptical argument, concerning deception.

The strictest proponents of the argument that enchantment is deception, self-illusion, and self-fraud are Robert and Linda Sparrow (Sparrow 2002, 2016; Sparrow and Sparrow 2006). They argue that most of the positive effects of the use of robots are the result of deception and that deception is a failure of the duty of human beings to perceive the world accurately; it is also immoral self-illusion and self-fraud since it involves believing that something is taking place when, in reality, it is not. Amanda Sharkey adds that deception might be seen as negatively affecting the human dignity of persons enchanted by robots, since it results in infantilization, in stimulating adults to behave like children who are not able to distinguish reality from fiction—this argument is particularly valid in the case of carebots and the elderly (Sharkey 2014). It needs to be emphasized that while the argument of deception appears most often in discussions concerning care robots, it is also present in examinations of sex robots—for instance, Florence Gildea and Kathleen Richardson claim that deception is involved even in the very expression "sex robots," since these robots do not provide sex (since, according to them, sex has to be based on mutuality, which robots cannot provide), but only masturbation (Richardson and Gildea 2017).

It can also be argued, however, that such enchantment and deception already occur, not only with intimacy robots, but on a regular basis. People tend to anthropomorphize cars, to laugh at fictional movies, to cry while reading fictional books. If there is nothing wrong with becoming enchanted by fictional tales, then there is also nothing wrong with becoming enchanted by robots—it is a natural phenomenon (Sharkey and Sharkey 2006). The deception argument is thus criticized in the following way: there is nothing wrong with people getting enchanted by robots, as long as they are aware that they are dealing with a machine. This is similar to people laughing and crying while watching a movie, knowing that they are dealing with fiction. People are not deceived as long as they are aware of the line between reality and fiction (Misselhorn et al. 2013; Sharkey and Sharkey 2006). Moreover, it is not clear whether the skeptical argument is right in assuming that most of the positive effects of interaction about intimacy robots derive from deception—some of them might take place even without the enchantment, that is: with a clear awareness that one is dealing with a robot, a mere object, an inanimate tool (Sharkey and Sharkey 2012).

There are also more radical (direct, or—in most cases—indirect) critiques of the deception argument. All the arguments described above assume that there is some boundary between objective (or at least intersubjective) reality, on the one hand, and illusionary, deceptive subjective experiences, on the other (although, of course, not all subjective experiences are deceptive and illusionary)—between the world as it really is and the way it appears. David Levy, however, consistently seems to ignore and delete this boundary. He asks, "if a robot *behaves* as though it has feelings, can we reasonably argue that it does not?" (Levy 2007a, p. 12). For Levy, the objective, internal properties are not in any sense more legitimate than the external appearance. He emphasizes, moreover, that these appearances do not even have to be objective or intersubjective—he suggests that subjective impressions are legitimate criteria for determining whether something is real or not. As he put it in one interview: "It just matters what you experience and perceive" (Choi 2008, p. 96). For Levy, therefore, the deception argument is wrong since there is no deception (at least in the sense in which the skeptics understand it) since subjective experience is a criterion of realness and it does not have to fit with any other inter-subjective or objective criteria—one cannot be deceived about one's subjective experience since subjective experiences are reality. Of course, it is easy to oppose this position from a so-called common sense point of view by pointing out that "love is blind" and people often tend to see love, beauty, or honesty where there is in fact none. Empirical studies that confirm this approach also exist (see Sullins 2012). The bottom line, however, is that Levy (supposedly indirectly and half-consciously) does not argue against common-sense beliefs, but rather tries to introduce a new common sense, to make a fundamental change in Western culture.

Regardless of which side is right in the debate over enchantment and deception, it is clear that this discussion extends beyond the area of discussions about robots and touches on a more general, ontological problem—what is real and will be real for individuals who participate in Western culture, and how the criteria of realness and thinking about realness might change—not only due to the robots, but also because of other technologies, particularly virtual reality (Boellstorff 2010). These broader perspectives are explicitly taken and articulated by Mark Coeckelbergh, who also links it with the debate on deception (Coeckelbergh 2012a) and David Gunkel—I will get back to this in Chapter 3. For now, I will stay with the (mainly) ethical axiological issues, at present focusing on those that concern social good, societal perspective, and effects on society.

2.3   Social Good, Societal Perspective, Effects on Society

As I mentioned earlier, when it comes to issues listed in the title of this section, intimacy robot skeptics are on the offense, in the sense that they initiate the discussion with their critical arguments while enthusiasts are on the defense with counter-arguments (while for individual good, individual perspective, and effect on individual, it was the opposite). Skeptics make the following two main arguments: first, that users of robots will become solipsists who objectify other human beings (eventually eroding the quality of social relations); second, that users will become imprisoned in their solipsism in the sense that they will be unable to participate in social relations (causing both the quality and quantity of social relations to decline).

2.3.1   Objectification, Solipsism and Isolation

The objectification argument is mostly articulated in reference to love-bots and sexbots and is mainly, though not solely, the work of feminist thinkers, especially Kathleen Richardson and Sinziana Gutiu. Simply put, objectification means that a (human) subject/person is being treated and/or perceived as an object/tool, that is: solely as a mean and not at all as an end, to paraphrase Kant's categorical imperative. To articulate how exactly interactions with lovebots and sexbots lead to the objectification of human beings (especially women), I would like to refer to Rae Langton's (2009) analytic work on the concept of sexual solipsism.

According to Langton, there are two kinds of sexual solipsism. The first one takes place when a thing or tool is treated as a person or subject in a sexual context—that is, when pornography is treated as a sexual partner, when it is used as a substitute for a human being. The second sexual solipsism concerns the opposite mechanism: when persons/subjects are treated as things/tools. In both cases, a solipsistic attitude is present in a different way: in the first case, the person who objectifies treats an object/tool as a person/subject, but in fact, only he is the real subject. In the second case, the person who objectifies knows that there are other subjects/persons around, but treats them as objects/tools, as if he were the only subject. Langton argues that these two types of solipsism are connected. The hypothesis she offers is that the solipsism of treating things (objects/tools) as people (subjects/persons) leads to the solipsism of treating people (subjects/persons) as things (objects/tools). Putting it more specifically, treating intimacy robots as persons/subjects leads to treating human beings (most often women) as tools/objects. This argument involves some assumptions and claims that need to be discussed.

At the very beginning, it is crucial to arrive at a definition of objectification that goes beyond the simple sense of "treating/perceiving a person/subject as a tool/object." To become objectified is to be deprived of autonomy and subjectivity. The objectified person is treated and perceived as if he or she has no beliefs and desires, no personality, no intentionality. The objectified person is thus nothing more than a means to someone's end and does not require any empathy, reciprocity, or recognition. I will now examine further details that are especially lively discussed by skeptics in their critical argument against objectification.

First of all, this argument assumes that intimacy robots will be treated as persons that do not have ability to consent (or not to consent). Sinziana Gutiu describes this as a "robotization of consent" (Gutiu 2012). It is closely connected with a fact emphasized by enthusiasts: that intimacy robots provide unlimited access to strong, new, positive experiences—this access is unlimited since robots do not say "no,"—nor, in fact, do they say "yes," since their consent is not important and in fact does not exist—robots are "persons" who do not and cannot say no. Obviously, skeptics argue that the "robotization of consent" will lead people who interact with intimacy robots to stop asking for consent not only from robots, but also from human beings and will thus objectify humans in that sense.

Second, as Richardson emphasizes, intimacy robots (especially sexbots) do not require any reciprocity or empathy. While in most cases intimate relationships between human beings there is a need and obligation for mutuality, this does not take place with intimacy robots. As we have already established, robots do not suffer and do not have feelings; they only *simulate* emotions to please their users. So, in fact, those robots do not require any love, care, or other form of intimacy. Again, skeptics worry that people who treat robots as commodities to consume, properties to possess, and tools to use that do not require any reciprocity or empathy will eventually treat human beings the same way (Richardson 2016a, b, c).

Third, lack of empathy and reciprocity together with the "robotization of consent" opens up a path to violence. According to Gutiu and Richardson, robots requiring neither consent nor empathy nor reciprocity encourage violent treatment that eventually may be shifted to human beings, especially women. Put simply, the use of intimacy robots may increase the level of violent behavior among humans in the intimate sphere, including rape, or at least result in desensitization (Whitby 2008).

In consequence, from a feminist perspective, love and sex robots reproduce and strengthen the stereotype of women as submissive tools that are "naturally programmed" to fulfill men's desires. They therefore strengthen gender inequalities, since they frame men as "real" subjects and women as submissive tools. Jennifer Robertson observes this tendency in Japan and calls it robo-sexism and posthuman sexism (Robertson 2010). Kathleen Richardson claims that the use of sex robots will lead all women to be seen as prostitutes: objects that cannot say no, submissive tools that do not require reciprocity, properties that can be used in any way, means to men's ends, humans who possess no subjectivity, no intentionality, and no autonomy that needs to be recognized (Richardson 2016b, c). It is also important to notice, as John Sullins and Neil McArthur point out, that intimacy robots may also inappropriately shape ideas about how the human body (should) look, e.g., reinforce the stereotypes concerning women's body and create pressure to achieve specific, exaggerated features of appearance that might eventually lead to an increase in eating disorders and other maladies (McArthur 2014; Sullins 2012).

It needs to be added, moreover, that the issues of objectification and solipsism are also closely connected with issues of replacement and isolation. Skeptics point out that solipsists who avoid dealing with other's people subjectivity by replacing them with robots or by objectifying them may want to do that, not in order to achieve some social-cultural skills, but permanently, because they prefer robots to people. In other words, many solipsists may share the enthusiasm of David Levy, who claims that lovebots and sexbots will be not only good enough to replace people, but will in some ways (ways that were broadly discussed in the previous section of this chapter) be better than people. For skeptics, such a decision leads to a self-isolation (in the documentary film "Guys and Dolls," one doll owner says, as previously mentioned, that he is like his own god living in his own world (Holt 2007)). It may eventually also lead to addiction. If people already tend to become addicted to pornography or video games, it is highly possible that the tendency toward addiction to intimacy robots, which may be considered better and more enjoyable not only in comparison to pornography or video games, but—according to some enthusiasts—also to human beings.

It is worth noting that skeptics aren't alone in raising this issue. Love- and sexbot enthusiast Neil McArthur also discusses the issue of isolation as a serious matter, calling it "cocooning." He believes that "cocooners" do harm both to themselves and to other human beings with whom they could potentially form intimate relationships (McArthur 2014). The choice of sexual solipsists to isolate themselves might also have serious additional consequences, similar to those suffered by people who are isolated against their will as a result of objectification. I will discuss these consequences in the next part of this section.

I would first like to examine the counter-arguments developed by enthusiasts of intimacy robots. The first one refers to the fact that there is no definitive proof that intimacy robots have, or will have, the negative effects that skeptics worry about. They emphasize the fact that skeptics' worries and doubts should be tested by empirical studies showing a causal link or at least a correlation between the use of robots and the tendency to objectification (Danaher et al. 2017). In other words, enthusiasts are optimistic not only about intimacy robots, but in general: they believe that unless we have a clear evidence that something is wrong, there is no need to worry. They argue that we need to know more about the impact of lovebots and sexbots before we decide whether or not to regulate or ban their use (Danaher et al. 2017). This is due to the fact that most enthusiasts seem to consider individual freedom to be a crucial value and they do not want to limit it to account for worries or doubts, only for decisive proof.

Enthusiasts also suggest that lovebots and sexbots may lead to exactly opposite results from those skeptics predict. They believe that lovebots and sexbots may decrease the level of objectification and violence in intimate relations. They assume that rather than stimulate desires for objectification and violent treatment, interactions with lovebots and sexbots will decrease it by working in a cathartic, therapeutic, compensating way (Danaher et al. 2017). These beliefs are based on studies of the impact of pornography that show the correlation between the availability of pornography and a decrease in rape and violent sexual behavior (see McArthur 2014). Skeptics nevertheless also base their own arguments on large volumes of empirical studies that deliver exactly opposite conclusions (see Gutiu 2012). The question of whether interaction with pornography, violent video games, or intimacy robots stimulates or compensates for negative attitudes remains to be solved, although it is not clear whether a solution can be delivered by quantitative empirical studies. Socio-cultural reality does not seem to be determined by laws like those that govern the natural, biological world. It is therefore hardly possible to establish a causal link. What can be established are only correlations, which provide rather weak arguments since even if we observe the positive correlation between "more or less pornography" and "more or less violence," we should be aware that there are plenty of other determinants that have an impact on the latter situation and, in fact, we do not know whether violence decreases or increases because of more or less pornography, or despite it.

This general problem, however, has an important philosophical background since it involves questions of the impact of fiction and fantasy on reality and the status of fiction and fantasy in general. It is worth mentioning that another context where this issue is present in the phenomenon of virtual realities such as "Second Life" (Boellstorff 2010). In this chapter, I have already discussed the issue of the boundaries between reality and appearance. The problem of the effects of fiction and fantasy, however, is a separate matter. It refers not only to the abovementioned dilemma of the impact of fiction on reality, but also about the reality of fiction. When we think of someone doing violence to lovebots, sexbots, or carebots that resemble humans but have no feelings, emotions, or consciousness, we may claim, as skeptics do, that it might have negative social effects, but most of us will also see nothing intrinsically wrong with such activity, since most people assume that one has a right to treat their (inanimate) property however one wishes. After all, such violence is a fiction, since no one is suffering. If we imagine that this lovebot or sexbot—or even sex doll—is in the form of a child, however, our attitudes may change. Such "fictional pedophilia" is, indeed, highly disturbing. Some scholars believe—more or less adamantly—that such a situation is unacceptable (Scheutz and Arnold 2016; Whitby 2008), while others—including John Danaher, a firm enthusiast of lovebots and sexbots—develop a whole argument against such performances (Danaher 2017). This is thus another part of the ethical and axiological debate over robots, in which the ontological question of what is real appears—it seems that Western culture has developed such sophisticated fictions that their status is increasingly less fictional and increasingly more real, or—to put it another way—the line between the real and the fictional becomes blurred.

Another counter-argument enthusiasts make involves the "nature" of technology. There is nothing inherent in the nature of lovebots and sexbots, they reason, that would lead to negative results such as objectification. That would be the fault of the people who would use robots in the wrong way, and if they do, it would be because there is something wrong with them or with society. We should therefore leave the robots alone to be used by people who do so properly, and focus solely on the individuals who are doing the objectification, since they are the source of the problem (Danaher et al. 2017). This point of view is based on a strong technological instrumentalism, that is, on the view that technology is a transparent tool, a neutral instrument that passively submits to the human will and has no impact or effect on its user or on the broader socio-cultural reality (except on those connected with the intentions of the tool users). Taking this point of view to an extreme, we might say that there is no point in debating robots (or any other tools) since there is nothing good or bad within them, and all the qualities and features of their usage are on the side of the user, and the usage of robots neither stimulate nor compensate either good or bad desires. A similar counter-argument points to the fact that there is no such single technology or any single factor in general that can profoundly reshape intimate human relationships (McArthur 2014). Although this might be true, skeptics may respond that robots are one kind of technology that is a part of broader socio-cultural tendencies that together might result in such a reshaping. Therefore, even if robots are only one form of technology and, as such, cannot on their own cause a significant increase in objectification, isolation, and solipsism, it needs to be acknowledged that there are already socio-cultural processes that result in tendencies to objectification, isolation, and solipsism, and robots may significantly increase their influence. I will discuss these processes and tendencies in the last section of this chapter.

There is a further counter-argument against the objectification thesis, one that particularly concerns the issue of consent. John Danaher claims that it is enough to program robots in such a way that they would be able to consent—to agree and disagree—to avoid the "robotization of consent" issue (Danaher 2017). Of course, skeptics may oppose this approach, objecting that, in the case of robots with no consciousness, no intentionality, no emotions, and so on, real consent is not possible, and only a simulation, an appearance of consent can take place. This in turn can be answered with David Levy's approach, discussed earlier in this chapter: "As to the question of a robot's being legally able to consent to its marriage, if it says that it consents and behaves in every way consistent with being a consenting adult, then it *does* consent" (Levy 2007a, p. 159). Put simply, perfect simulation and persuasive-enough appearance are as real as reality, so the problem of the "robotization of consent" disappears with the appearance of appropriately sophisticated robots.

Finally, not only do enthusiasts try to argue against the risk of objectification and other negative phenomena, they also claim that the use of lovebots and carebots will improve the quality of human-human interactions, particularly intimate ones, and help to shape new kinds of relationships (rather than degrading them through objectification and other negative phenomena). Enthusiasts claim that lovebots and sexbots will increase the amount of pleasure in already existing human-human intimate relationships. They believe that lovebots and sexbots will educate humans and thereby improve their (technical) sexual skills, enable them to escape routine, and help in cases of an imbalance of libido within the relationship, so that the partner with lower libido would not be constantly prompted to intercourse while the partner with higher libido would be able to satisfy his or her needs without having to resort to cheating (McArthur 2014). Lovebots and sexbots would thus stabilize existing relationships by making them more diverse and more pleasurable and by decreasing the risk of infidelity or breakup as a result (Levy 2007a, pp. 307–308).

As for helping to shape new intimate relationships with people, enthusiasts claim that relations with robots might be therapeutic and a practice (as discussed earlier) that enables people to overcome personal obstacles (shyness, for example) and to obtain socio-cultural interaction skills that will enable them to engage in intimate relationships with human beings (McArthur 2014). Therefore, from this point of view robots not only will not lead to isolation, but also help to escape from it. This enthusiastic approach is, of course, vulnerable to criticism due to certain silent assumptions on which it is based. One assumption pro-robot thinkers make is that both participants in an intimate relationship will agree on the use of intimacy robots and that neither of them will consider it to be an act of infidelity nor will they feel jealous—this assumption seems to be highly controversial. Scholars who discuss these issues (Borenstein and Arkin 2016; Yeoman and Mars 2012) do not provide any definite answer. On the one hand, sex with a robot could be considered infidelity, if the robot is perceived and treated as a subject/person. On the other hand, if robots are perceived and treated as a tool/object, then sex with them might be considered to be no different than, for example, using a vibrator. It therefore might be a matter of subjective experience and subjective opinion (although one can easily imagine how sooner or later intersubjective cultural patterns may develop, making the evaluation of such phenomenon far less based on a subjective approach). Nevertheless, once more, I would like to point out that similar issues are connected with intimate relationships in Second Life (e.g., can a marriage in Second Life be a cause of jealousy or divorce in the "first life"?) (Boellstorff 2010). A second rebuttal to enthusiasts is that they assume that a scenario in which people become solipsists who replace human companions with robots will not take place; enthusiasts imagine only that robots will be supplements to intimate human relationships, not that they will be substitutes for them. In other words, their argument expresses disbelieving in the possibility of solipsism, replacement, and isolation.

Moreover, enthusiasts' therapeutic argument does not square well with another claim enthusiasts make, emphasized particularly by David Levy, that robots will be—in some aspects—better than people. If so, why would anyone bother dealing with people if robots are easier and better? (Borenstein and Arkin 2016). Enthusiasts' arguments therefore mostly assume that eventually people will perceive and treat robots as supplements, assistants, and means for other ends, rather than as substitutes, replacements, and ends in themselves—although, at the same time, they claim that robots can replace people in the sense of providing what people do provide in intimate relationships in a better version and without effort. This part of the enthusiasts' argument thus appears incoherent. I can only speculate about the enthusiasts' possible answer to the risks connected with the replacement scenario. Perhaps they would say that such a scenario will not take place since robots and humans are different, and therefore humans needs can be best covered when one interacts with both humans and robots, since none of them can provide everything one needs or wants.

Enthusiasts might also say—given their emphasis on individual freedom and enjoyment—that replacement, solipsism and isolation are not problems at all. Self-isolated solipsists may make a free choice to replace people with robots. If they are happy with this choice, then there is no problem at all. If they are unhappy, they should make another, better choice. I base this speculation on their axiological and ontological assumptions about the value of individual freedom and the human condition. I will discuss those assumptions in more depth in the last section of this chapter. In the next part, I will discuss the skeptical argument that involves the further negative consequences of self-isolation and solipsism that also seem to pose a stark contrast with enthusiasts' general assumptions. But first, I will discuss the problem of objectification and isolation specifically as it relates to carebots and the elderly.

The skeptical argument about objectification and isolation pertains not only to lovebots and sexbots, but also to carebots and artificial companions for the elderly. Here, objectification also means treating a person/subject as a thing/object. The objectified person is treated and perceived as if he or she has no beliefs, desires, or intentionality and, in consequence, the objectified person is nothing more than a means to someone's ends and requires no empathy, reciprocity or recognition.

The main way such objectification may take place is by considering elderly people to be nothing more than problems that need to be solved by technological means, as objects that need some technical maintenance and that do not require any form of social recognition, empathy, or respect (Sharkey and Sharkey 2012; Sparrow 2002, 2016; Sparrow and Sparrow 2006). The worry therefore concerns the risk of developing a reductionist view of the needs of elderly people and of the care they require.

Scholars have conceptualized this reduction of care and needs in two related ways. Parks suggests that robots will only care *for* the elderly (by fulfilling their physical needs), but will not care *about* them (by providing attention, love, sympathy, recognition). Coeckelbergh makes a similar distinction between deep care and shallow care (Coeckelbergh 2010). Borenstein and Pearson (2010) make the same distinction. There is a general agreement that in the near future carebots will be able to provide only shallow care and to care for elders, but not to care about them. Most scholars therefore argue that carebots should be a supplement or an addition to human care, not its substitute or replacement. Some suggest that such automated assistance may eventually improve standards of care, since human caregivers will not have to deal with technical, arduous, and often unpleasant tasks (concerning hygienic matters, for example), enabling them to focus their efforts on providing deep care and caring *about* their care recipients (Borenstein and Pearson 2011; Pfadenhauer and Dukat 2015). This may empower the elderly to the extent that they would not be forced to leave their homes and move to public or private facilities that provide care (assuming that elders will be able to afford such carebots).

This moderately optimistic scenario has been criticized by more radical skeptics, such as Robert and Linda Sparrow, who believe that carebots will not supplement and assist human care, but rather replace and be a substitute for it (Sparrow 2016; Sparrow and Sparrow 2006). Their view is based on the belief that the quality of elderly care is already low and declining, since the elderly are increasingly seen as objects that require technical maintenance, and the main problem of the system of care is not how to provide deep care for the elderly, but how to cut costs on the care they receive. The Sparrows therefore suggest that robots are likely to become cheaper, low-cost replacements for human caregivers and the needs of elders will be adjusted to carebots skills, not the other way around. Sparrow claims that this is an example of searching for "technological fix to a social problem" (Sparrow 2002, p. 18), which is not technical problem at all. Note that the Sparrows do not claim that it is anyone's intention—least of all that of the roboticists and other engineers who develop robots—to produce carebots that will eventually replace human caregivers. What they claim is that such replacement (resulting in the objectification of elders) will be an unintended consequence of the implementation of carebots, a development determined by structural factors, particularly by economic pressures to cut costs (especially in social care) and the already-existing social tendency to objectify the elderly in more or less direct ways by having a minimalist conception of their needs. Similar prognoses are suggested by Parks and Misselhorn (Misselhorn et al. 2013; Parks 2010). The Sparrows worry that carebots will replace human caregivers, both in eldercare facilities—due to the tendency to cut costs—and in their homes—where the visit of a caregiver and/or a cleaner is a rare chance for the elderly to have contact with other human beings and enjoy a bit of "human touch."

The risk of objectification, in the case of carebots and elders—as well as in the case of lovebots and sexbots—is closely connected with the risk of isolation. If robots replace human caregivers, care recipients will be deprived of "human touch" and human contact (Borenstein and Pearson 2010; Sharkey and Sharkey 2012; Sparrow 2016; Whitby 2011). Some studies, on the other hand, suggest that robots may help mitigate isolation. The robot Paro, for example, has stimulated and enhanced social contact among patients with dementia in eldercare facilities (Sharkey and Wood 2014; Sharkey and Sharkey 2010a). Studies also show that autistic children's social skills improve when they interact with artificial companions (Borenstein and Pearson 2010; Robins et al. 2005; Scassellati 2007; for a detailed investigation of the issues of autism and robots see Richardson 2016d, 2018). It must be emphasized, however, that in both cases robots were only supplementing and assisting the caregivers and therapists, not substituting for and replacing them.

To summarize, the main dilemma is whether intimacy robots (e.g., sexbots and carebots) will replace and substitute for or assist and supplement human contact. The former possibility is seen as leading to the objectification of human beings (especially of women and the elderly) and to a degradation in the quality and/or the presence of intimate relationships and care. A blend of robot and human relations, on the other hand, is perceived as a chance for increasing the level of the quality of intimate relationships and care. Shannon Vallor aptly puts it this way: care robots may either liberate people (particularly caregivers) *from* care or liberate them *to* care (Vallor 2011). I would suggest that this maxim can be generalized to all intimacy robots, who might liberate us either *from* being with other people or *to* be with other people. The way it will go depends on how robots are designed, on individual choices, and—on the macro level—on dominant cultural values and beliefs as well as social and structural conditions that determine people's choices. These factors will decide whether and how intimacy robots will enchant us and change our vision of humanity and human needs. I return to this issue more broadly in the last section of this chapter.

2.3.2   Solipsism as Imprisonment

I have discussed the issue of objectification, isolation, and solipsism. Solipsists are people who (for reasons I discuss later in the chapter) do not want to deal with other people's subjectivity and to avoid it replace people with robots (thus substituting subjects with objects) or objectify people (treat and/or perceive subject as object). Objectified people are those who are treated and/or perceived as objects and are left alone with robots or with people who treat them as things. Both solipsists and the objectified are isolated and deprived of sharing "human touch": social connectedness, symbolic recognition, empathy—solipsists willfully deprive themselves of "human touch" by their own choice—though if they choose to replace people with robots rather than objectifying people they also become deprived of "human touch." The objectified are likewise deprived of "human touch" (by the solipsists' and objectifiers' choice). I would like to show that this state of affairs may radicalize itself into a state of being imprisonment in solipsism that refers to a cultural "state of mind."

The main difference between the "regular" solipsist and the objectified person, on the one hand, and the imprisoned solipsist, on the other, is that while the former are not involved in human relationships because they do not want to or have no opportunity to do that even though they are able to do that, the latter is unable to conduct a human relationship even if he or she wants to and has an opportunity. Obviously, the above statement is based upon a strong assumption about the cultural character of the social skills that enable people to "share" the "human touch." I would suggest that no one is born with hard-wired social skills (although I agree that we are born with disposition to achieve them in the process of socialization and enculturation). Rather, we have to learn them (disturbing cases of children raised in isolation seem to offer enough proof of this). What is more, since they are learned, it might also be possible to unlearn and lose them. Solipsism as imprisonment, in the sense that I propose, is an effect of losing social and cultural interaction skills of meaningfully communicating and emotionally empathizing that enable people to have human relationships. Therefore, whereas in the earlier part of the chapter the main worry was that people may exclude themselves from sharing "human touch" (by replacing people with robots and objectifying people) and other people will be excluded from this sharing (by being treated/perceived as objects or left alone with robots), here, the main worry is that those people will lose their ability to participate in this sharing no matter whether they want to or have opportunity.

This skeptical argument may sound extreme and over the top, yet it is solidly grounded in studies conducted by Sherry Turkle (2012, 2015). Turkle's research relates not only to intimacy robots, but also to other technologies, especially to those that mediate communication. Turkle writes that in her early work she believed that technology and robots may become a form of practice that eventually would improve the social and cultural skills that enable people to have meaningful intimate relationships, and that it will be possible to move and implement those skills, once developed, into authentic, non-mediated face-to-face human interaction. Nevertheless, following years of research, she has changed her mind and now feels that interactions with technology or mediated by technology do not improve direct, face-to-face interactions among people. People tend to use technology not to obtain or to improve skills necessary for face-to-face interactions with people, but rather as compensation for the lack of these interactions—they replace direct, face-to-face interactions with interactions mediated by technology or with technology. People who have recourse to technology consequently not only fail to increase their social skills, but they also lose them. The starkest examples of this are children interviewed by Turkle, who explicitly say that they would like to be able to participate in human-to-human, face-to-face interaction, but all they are have learned and are able to do is to chat online and send text messages. Turkle argues that this lack of ability to communicate is closely connected with a lack of a more general skill: empathy. She finds that people who tend to use technology to mediate their communication not only lose the skill of developing their own communications in real time, but also have difficulty understanding the communications they receive—especially non-verbal communications—as well as the ability to imagine themselves in someone else's shoes. This issue is particularly visible in children, since their social and cultural interaction skills are not yet fully developed and remain fragile.

For now, the discussion has touched less on the problem of carebots for children (robot nannies) and robot teachers than issues connected with intimacy robots for adults and the elderly, although some examinations do consider comparable problems (Borenstein and Pearson 2013; Pearson and Borenstein 2013, 2014; Sharkey 2016; Sharkey and Sharkey 2010b; Whitby 2010). The attitudes in these discussions are far more moderated and careful than in discussions of other intimacy robots, not only because there are fewer empirical studies, but also because children are obviously more fragile and vulnerable to robots' potential impact. The separation between enthusiasts and skeptics is therefore less clear and researchers tend to present considerations of both the opportunities and the risks of introducing robot nannies and robot teachers. In fact, both opportunities and risks are comparable to those discussed in the case of other intimacy robots. Of course, there is an important difference. With adults and the elderly, the robots are interacting with "developed," "mature" human beings, who—in most cases—already possess some socio-cultural skills. With children, on the other hand, the people the robots interact with is "not-yet-developed" and "immature" in the sense that socio-cultural skills are not yet in place. Considering children in this context thus demands for more painstaking attention. Still, discussions of robot nannies and teachers do not involve any new issues, at least none that are significant for my approach (although, let me make clear that I consider issues connected with robot nannies and robot teachers to be far more important than the abstract divagations that I propose).

I will now discuss possible rebuttals to the skeptical argument about being imprisoned in solipsism. First, it needs to be emphasized that solipsism as imprisonment may appear only if it is preceded by "regular" solipsism, objectification, and isolation—and according to enthusiasts, these issues are very unlikely to take place. So, if there is no need to worry about the solipsism, objectification, and isolation, then there is also no need to worry about being imprisoned in solipsism.

Second, as previously mentioned empirical studies show, robots might stimulate and improve human interaction skills and foster social interaction, for example, in clients in eldercare facilities and in children with medical autism. Such studies are supported by a more speculative argument provided by Eleanor Sandry, who argues that "non-humanoid robots might form new kinds of relations with humans, which do not need to be framed as mirroring human-animal or human-human interactions. The companionship of such robots can therefore be figured as something new and different, not as a replacement for existing relations, but rather as an addition" (Sandry 2015, p. 95). According to Sandry, not only will human-robot relationships not replace relationships among humans, but they will increase the quality of interhuman relations by teaching human beings communication that is not based on eliminating Otherness, but on recognizing and celebrating the Otherness of the Other: first, that of robots, then that of fellow human beings.

Again, it is crucial to point out that this sort of counter-argument implicitly assumes that robots will be perceived and treated as supplements and additions and not as substitutes and replacements. It is simply not clear whether this assumption is valid, and some empirical evidence—which will be thoroughly discussed in the next chapter—suggests that people have a tendency to treat robots more as subjects and persons than as tools and objects, which increases the chance that they will replace human persons (though, of course, robots persons can still be supplementary to human persons, or—to put it another way—human persons and robots persons may be complementary. Indeed, as we saw earlier with David Levy, some enthusiasts claim that there is no difference between human beings and robots who persuasively simulate human beings. Ascribing such status to robots, they seem to suggest that they could potentially replace human beings. It is also possible, however, that a robot or other object, perceived or treated as a tool, could also replace human being, though the possibility seems more remote).

I believe that there are two possible answers to this question from an enthusiasts' perspective. The first involves David Levy's belief that there is nothing to lose if we replace robots with humans. If, as Levy claims, there is no difference between perfect robot simulation and real humans, then even the "human touch" can be simulated, and the only difficulty here is the technical one of building robots that are sophisticated enough to simulate people perfectly and appear utterly as human beings. The second response might acknowledge that robots—even those that perfectly simulate human beings—do not provide people with "human touch," but that is nothing to worry about, either, since people do not actually need "human touch" at all. Both of these answers lead us back to general problems of the boundary between reality and appearance, but also to the more general question of the status of humans and of human needs—particularly of the need for "human touch"—and also of the status of robots. I will discuss these issues in the next section of this chapter.

2.4   KEY ISSUES

2.4.1   Axiological Issues

I will begin by examining the axiological disagreements between enthusiasts and skeptics. I believe that it is crucial to understand what values—implicitly assumed or explicitly stated—underlie both enthusiasts' and skeptics' positions—in that sense, I point to the rather obvious fact that the groups differ not only when it comes to facts, but also when it comes to values. Enthusiasts are—by definition—more optimistic and open to intimacy robots and less interested in, or at least less worried about, potential risks (these features are very closely connected with their ontological beliefs about humanity that will be discussed below). It seems that the crucial value for them is individual freedom, particularly freedom of choice. Skeptics are—by definition as well—more careful and reserved, as well as more worried about potential risks (again: it is deeply intertwined with their ontological beliefs) and their crucial value seems to be the quality of human-human interactions that they consider to be a most important element of common good, social well-being (and eventually probably also individual well-being as well). It is clear that enthusiasts focus on the individual good and on the situation with(in) the individual, while skeptics are primarily concerned with common/social good and on the situation among individuals. Obviously, this is why enthusiasts are on the "offense" when it comes to issues connected with individual goods and effects on individuals, while skeptics are on the "offense" when issues concerned with societal good and effects on society and quality of interactions are involved.

Moreover, I believe that it is worth connecting the issue of the approach towards freedom of choice with the attitude towards the contemporary capitalistic market and consumerism. Enthusiasts seem to tend to believe that thanks to the capitalistic market we can choose from the broad spectrum of commodities and that the economy should not be regulated by the state or any other institutions since that would limit the spectrum of choice. Therefore, I would say that enthusiasts are implicitly in favour of the market values, economical growth and capitalistic order in general as something positive. On the other hand, skeptics seems to be more reserved about the role of capitalism and the function of the market. They do not see the market simply as a place where people meet to exchange goods basing on their freedom of choice, but rather consider it a structure that has an impact on human beings and social relationships, particularly by reducing the human relationships to market transactions. Therefore, they assume that economic growth is not simply an unproblematic positive value, since it might entail serious non-economic costs. Moreover, at the personal level the hope for individual economic profit might result in a conflict of interest e.g. when someone who invests in new technologies, particularly robots, at the same time develops arguments in favour of these devices and presenting them as a neutral academic scholarship.

It is worth embedding this issue in the broader cultural context, since similar axiological tension regarding individual freedom and the function of the capitalistic market has already been developed in reference to contemporary intimate relationships without the presence of robots at all. On the one hand, Anthony Giddens (1992), Jeffrey Weeks (2007), and Brian McNair (2012) have noted an increase in the spectrum of individual choices and emphasize the positive opportunities for individuals that can result from it, remaining uncritical towards any kind of a possible impact of the capitalistic market. On the other hand, Urlich Beck and Elisabeth Beck-Gernsheim (1995), Zygmunt Bauman (2003), Arlie Russell Hochschild (2003, 2013), Eva Illouz (2007, 2012), and Wojciech Klimczyk (2008) all emphasize a decrease in the quality of intimate relationships as a result of their rationalization, commercialization, and commodification, which in a large degree is determined by the capitalistic market. It is worth to mention that this is strictly connected with methodological differences between methodological individualism typical of enthusiast and methodological anti-individualism typical of skeptics—these two positions are discussed in the next section.

As in the debate examined earlier, enthusiasts point to increasing freedom and positive experiences (but not security), while skeptics emphasize objectification and solipsism. It therefore needs to be acknowledged that most of the issues that concern intimacy robots were already being discussed before these robots even appeared, since they concern more general socio-cultural tendencies. It can be also said that the development of intimacy robots is a result of those tendencies, and that intimacy robots will, in turn, increase the significance and influence of those tendencies. This is especially visible for carebots. Scholars emphasize that they are being developed as a response both to a demographic situation and to the general devaluation of care itself, which is viewed as a burden, an unpaid "women's task" or a low-paid job (Hochschild and Ehrenreich 2013; Parks 2010; Vallor 2011). This devaluation results in low-quality care and its rationalization, commercialization, and commodification (as with other intimate relationships), particularly in public facilities (Sparrow and Sparrow 2006; Vallor 2011). Mark Coeckelbergh claims—following Max Weber and other classic theorists of modernity— that many of those issues and tendencies are internal, "natural," and almost inevitable elements of modernity (Coeckelbergh 2015). It therefore needs to be emphasized that intimate relationships with humans are increasingly often seen as a burden. Love and care are something that people want to receive, but not to provide, while the human subjectivity of other people becomes an obstacle that limit's individuals freedom (Melson 2010). The tendency to disenchant human beings (e.g., the sentiment that there is nothing extraordinary or magical about them, that their character is fundamentally problematic) was present before robots, and robots—and the fact that they are enchanting humans—seem to be one of its consequences.

Nevertheless, robots significantly shift the "axiological balance" in discussions of intimate relationships. The enthusiasts' argument, on the one hand, is in a strong position, since in the case of intimate relationships with robots, freedom, and positive experiences are accompanied by increased ontological security, which was highly precarious in the "pure relationships" celebrated by Giddens, which are temporary by definition. Users of intimacy robots therefore obtain not only freedom and positive experiences (already available in human-human "pure relationships"), but also the security that results from total control over the robot—such control is not (legally, at least) available over a human being, neither in pure relationships nor in objectifying relationships. On the other hand, this shift also adds something to skeptics' arguments that call attention to the increased level of solipsism, isolation, and risk of further objectification, as well as to the fact that robots are not able to fulfill the second-order desires.

I would therefore like to argue that enthusiasts express and contribute to the growing tendency to disenchant humans and intimate relationships among people. First, their claims express and result in a disenchanting *of* humans and their intimate relationships, particularly when they say that there is nothing extraordinary or "magical" about either humans or their intimate relationships, nothing that could not be rationally calculated, engineered, and implemented with robots—McArthur (2014), for example, claims that sex is just like any other bodily pleasure, nothing more than some technical action that involves physical processes stimulating neurons in the brain. Second, their positions express disenchantment *with* humans and their intimate relationships: enthusiasts suggest (more or less explicitly) that humans and interactions with them are often difficult, messy, painful, and disappointing and that robots offer a better alternative, since robots offer not only more extreme positive experiences and higher degrees of freedom, but also ontological security—the sense of total control, of achieving everything one desires without any unnecessary effort. For enthusiasts, the case is easy: why bother with people, if robots have all the good stuff in a better and more easily accessible version, and no bad stuff? People and the relationships among them are becoming disenchanted (there is nothing special about them except the unnecessary obstacles, limitations, and disappointments) and this is why robots can be enchanting (they give everything people can give, but more of it and without the drawbacks).

On the other hand, skeptics oppose the disenchantment of people and relationships among them and criticize the enchanting character of robots as deception. Skeptics believe that there is something about interactions with people that cannot be present in interactions with robots: recognition, desire for desire, understanding, and empathy: the "human touch." For enthusiasts, of course, all of valuable features of human-human interaction can be simulated by robots. For skeptics, even if that is true, simulation will never provide the same experience as the real thing—even if the experiencing subject claims to feel no difference. Skeptics oppose the disenchanting human-human interaction by claiming that their quality, authenticity, and other features cannot be engineered, technically achieved, or simulated. Although they admit that the disenchantment of humans and their intimate relationships actually does take place quite often, they claim that it is important to try harder to find enchantments in people rather than to seek false—in their opinion— enchantments in robots.

Skeptics thus do not find robots enchanting, but rather deceiving, creating an illusion that they offer and provide something they do not have at all. Again, it needs to be pointed out that this disenchanting tendency is a broader and earlier phenomenon than the appearance of intimacy robots: various scholars have suggested that intimate relationships between humans are becoming rationalized (Illouz 2007, 2012), commercialized (Hochschild 1983, 2003, 2012), commodified (Bauman 2003) and that some disciplines—particularly economics—suggest that all human interactions (including intimate relationships) are nothing more than market transactions (Michael Sandel 2012). Moreover, Illouz, Hochschild, and Bauman show how intimate human relationships are increasingly becoming a source of disenchantment and disappointment.

There is one more remark about axiological issues I would like to make. Enthusiasts—such as Levy—seem to believe that individuals know what is best for them, and it is always good to provide individuals with easier access to what they want. But skeptics—such as Sullins and Turkle—tend to suggest that what you want is not always what you need or what is best for you, and that sometimes the easier way is not better than the more difficult one. On the one hand, therefore, we have an optimistic vision of people who know what is best for them so we need to help them to achieve their aims more easily and effectively, and on the other, we are suggested that people might be wrong about what is best for them, especially when it comes to the claim that easier is always better. This inevitably leads to the question of how to evaluate what one needs and whether his or her needs are satisfied—the following section examines this and other methodological issues in more details.

2.4.2   Epistemological and Methodological Issues

I would now like to discuss the epistemological and methodological level, where at least two matters need to be examined: determining factors and measurement. It is rather clear that enthusiasts are methodological individualists: they believe that the crucial (or maybe even: the only) factor that shapes transformations of socio-cultural reality is the individual subject and his or her intentions and actions. Skeptics, on the other hand, believe that also (or maybe even: above all) objective, non-individual factors need to be concerned when it comes to explaining and predicting transformations of socio-cultural reality. This is why

David Levy, John Danaher, Neil McArthur, and Robin McKenzie believe that increasing the spectrum of individual choices is a good thing, since sooner or later individuals will choose from this spectrum what is best for them, and if all individuals do that, it will also be good for society. And this is why Kathleen Richardson, Sinziana Gutiu, Sherry Turkle, Robert Sparrow, and Linda Sparrow believe that "good intentions" are not enough to be sure that everything will be all right, since human intentions and actions are (more or less directly) shaped by various structural factors, and moreover, human actions lead not only to intended, but also to unintended consequences.

We might again think back to the discussions about intimacy "before and without robots." On the one hand, enthusiasts such as Anthony Giddens, Brian McNair, and Jeffrey Weeks have celebrated increasing freedom of individual choice and "reflexive" attitudes of individuals while almost entirely ignoring non-individual, structural factors. Meanwhile, skeptics such as Urlich Beck, Elisabeth Beck-Gernsheim, Zygmunt Bauman, Arlie Russell Hochschild, Eva Illouz, and Wojciech Klimczyk have emphasized that free and reflexive choices are determined by strong structural factors, especially the tendencies of contemporary capitalism such as commercialization, commodification, and consumerism. In fact, the situation in discussions about intimate relationships with robots seems to be comparable. On the one hand, enthusiasts do not consider structural factors, but rather focus on increasing the spectrum of individual choice and the opportunities for individuals that may result from it. Alternatively, they focus on abstract analysis—typical of analytical philosophy—concerning the lack of proof for causality or correlation between robots and pornography and objectification, and so on. Skeptics, on the other hand, point to prevailing contemporary sociocultural tendencies concerning objectification, "regular" solipsism, isolation, and solipsism as imprisonment (especially those connected with pornography and mediating technologies), as well as to the structural factors that determine them—neoliberal capitalism in particular (Richardson 2016b). For the former, everything seems to depend on individual choice and good intentions are assumed to lead to equally good consequences. For the latter, individual choice depends on structural factors and may lead to unintended consequences.

A second methodological problem, which has not appeared at all in the foregoing discussion, is still significant both in general and also as a part of my further considerations. It concerns the measurement of well-being. It is not clear how exactly we should determine whether robot users are really (whatever this "really" means) better off or worse off. Obviously this issue is not a new problem, but there are some aspects of it connected with intimacy robots that I would like to examine, since they relate to other issues that I discuss here.

On the one hand, Robert Sparrow and Mark Coeckelbergh suggest that we need some objective (intersubjective) measures of human well-being, and, on the other, David Levy suggests that the proper measure is the subjective experience of each individual. Of course, as Coeckelbergh points out (2010), objections can be made to both of these approaches. The subjective approach can be criticized because people may subjectively feel—as result of propaganda or brainwashing, and so on—that everything is great, when in fact their life may be horrible. The objective approach, on the other hand, can be criticized as a kind of paternalism that treats people like irresponsible children, limits their freedom, does not consider their subjective experiences and opinions and "knows best" what is good for them, not mentioning the fact that people are different (both across different cultures and within each culture), and their needs may also vary significantly and are subject to change (Sorell and Draper 2014).

The reference point chosen to evaluate the results of measuring is a key factor. As Sparrow (2016) remarks, the same effects of robots may be interpreted differently depending on whether we will compare them with—for example—bad practices in public caregiving (the results will then come as an significant improvement) or with the care provided by devoted family members (the results will then appear inferior). It is therefore important to be aware of the various issues connected with the interpretation of empirical qualitative studies, not to mention the fact that some of them are funded by robot producers, which raises doubts about their reliability (Sharkey and Wood 2014). More generally, the discussion of whether intimacy robots are good or bad is not only about focusing on different values or pointing out different facts, but also about arriving at different interpretations of the very same facts.

Therefore, it is crucial to notice that in the case of both methodological issues enthusiasts' approach is based on individual and his or her subjective experience, while skeptics' approach is a result of focusing on (but not limiting to) objective (intersubjective) structural conditions. It is clearly compatible with axiological positions of both sides. To summarize we might say that in reference to both axiology and methodology enthusiasts are focused on individual and his or her subjective choices and experiences, while skeptics are primarily concerned with the objective (or intersubjective) quality of interactions between individuals as well as factors that determine them.

2.4.3   Ontological Issues

The main ontological issues in the discussions about intimacy robots are the problems of the boundary between reality and appearance on the one hand, and the boundary between human beings and robots, on the other. These problems are tightly intertwined, of course, but I will discuss them one after another to deliver the conclusion that robots can enchant us mainly due to the fact that we are disenchanting human beings.

The problem of the boundary between reality and appearance has been present throughout this whole chapter. I would argue that this boundary is currently quite fluid and its status is under discussion. While skeptics would like to leave it as it is—between objective/intersubjective reality and subjective illusion—enthusiasts would like to shift it or, preferably, delete it (by saying it has never really existed and claiming that what is a part of subjective experience is as real as anything that is considered objectively or intersubjectively real). As I have already mentioned, this issue is not only a side effect of enthusiasts arguments, but also a matter of detailed debate, especially as a part of the "relational turn" proposed by David Gunkel and Mark Coeckelbergh (it is important to distinguish this particular notion of relational turn from many others that have been discussed in humanities and social sciences—see [Dépelteau 2013] for an analysis of various relational turns). I will examine their point of view in Chapter 3, since I see their approach as going beyond the question of "how should we think about robots and humans (and appearance and reality)?" to the meta-level from which they ask "how should we think about the way we think about robots and humans (and appearance and reality)?"

The second ontological issue concerns the status of human beings and robots and the boundary between them—obviously, these ontological issues have also strong axiological implications. I would like to point out that there are two grounds on which the discussion about the boundary takes place. The first is the obvious one between the enthusiasts and skeptics with whom we are already familiar. The second, however, is between other parties, whom we may label as "non-anthropocentric posthumanists," on the one hand, and "anthropocentric humanists," on the other.

As for the first ground, enthusiasts are rather undecided about the status of robots, though they are clearer concerning status of humans. As for the status of robots, enthusiasts are undecided in the sense that when it is convenient they claim that robots are subjects/persons, and when they need to strengthen another argument they consider robots to be objects/tools. (Some enthusiasts are far clearer in that regard. John Danaher and Neil McArthur, for example, seem to view sex robots as tools/objects.) The best example is David Levy, who claims that robots are like vibrators and that we shouldn't ban the former if we do not ban the latter (Levy 2007b), but also that states that he sees "differences between robots and humans as being no greater than the cultural differences between peoples from different countries or even from the different parts of the same country" (Levy 2007a, p. 112), and that banning marriage with a robot is like banning homosexual marriage (Levy 2007a, p. 304). We could speculate that in the former case, Levy is talking about unconscious robots, and in the latter about conscious one. For Levy, however—in accordance with his general claims about the relation between the real and the apparent—a robot that *appears* to be conscious *is* conscious (Levy 2009). As for the status of humans, it seems that enthusiasts see human beings naturalistically, as a system of chemical and biological processes and features. This is why they often make reference to naturalistic psychological research (as Levy does) or claim that sexual pleasure is simply a bodily pleasure like any other (as McArthur does).

Enthusiasts thus disenchant not only intimate relationships between human beings and human interactions in general, but also the status of the human being itself—there is nothing about human beings that cannot be engineered, there is nothing except pure physicality, chemistry, and biology. This aspect of the disenchantment of humans is therefore strongly connected with tendencies to naturalize humanity, to suggest that everything that happens in humans and with humans is nothing more than a kind of chemical and biological process. As a consequence, everything that is going on with(in) human beings can be counted and calculated and therefore engineered into robots. Of course, this is not a new idea and a lot of various disenchanting approaches toward human beings have been developed previously.

Obviously, skeptics adamantly oppose this approach. First of all, they oppose the claim that robots are subjects/persons—they do agree that some people in some circumstances may tend to treat robots as subjects/persons and that robots may perfectly simulate and persuasively appear as subjects/persons, but it does not make them subjects/persons at all, and should be simply understood as a mistaken perception (and that, in fact, is based on the previously mentioned assumption about the existence of a clear boundary between reality and appearance). They claim that robots—from the objective/intersubjective point of view—are tools/objects and they should be treated as such—Joanna J. Bryson has offered one of the most influential positions that express such an approach toward robots (Bryson 2010; Bryson and Kime 2011).

As for the second ground, which in fact brings us back to an axiological level, non-anthropocentric posthumanists believe that considering people to be more important and better than the rest of the world is bad both for the rest of the world (mainly for ecological reasons) and to humans (who impoverish their own experience of reality by treating everything except of themselves as objects/tools). Non-anthropocentric posthumanists therefore consider robots and interaction with robots an opportunity to blur the boundary between the human and the non-human to make people more sensitive to the Otherness and thereby enrich their experience of reality (Sandry 2015). This is why thinkers like David Gunkel and Mark Coeckelbergh try to open alternative paths of thinking, including non-modern thinking, of going beyond the anthropocentric humanism—Gunkel (2018) is particularly known for his examinations of the idea of robot rights, which in fact are no longer only a speculation since David Hanson's robot Sophia has been granted with citizenship of Saudi Arabia in 2017, February 27, 2018 has been officially proclaimed robot Sophia's day in the state of Minnesota and AlphaGo's victory over the human champions in ancient game go has been celebrated by the monument in the center of Vienna (see the next chapter for more details on Coeckelbergh's and Gunkel's approach). From the perspective of anthropocentric humanists, the main critical doubts about non-anthropocentric posthumanism and its proponents seem to be about the necessity and consequences of implementing their ideas. On implementation, Anne Gerdes (2016), drawing on Kant, tries to show that we do not need to become non-anthropocentric posthumanists to care for animals, the environment or even robots, since it is possible to express the necessity and importance of the well-being of Others and of Otherness on the grounds of anthropocentric and humanistic ethics. As for the consequences, Kathleen Richardson (2015, 2016a, b) worries that making humans and non-humans more equal participants of a harmonic world that does not belong solely to humans, but to all organisms, may be achieved not by elevating other organisms to the status of humans, but by degrading humans to the (contemporary) status of non-humans, and that this will result in perceiving and treating everything—both humans and non-humans—as tools and objects, which would not be good for either category. In other words, Richardson worries about strengthening the tendency that Charles Harvey describes, talking not only about the impact of sex robots but also the impact of the internet on human mentality: "As machines become more human-like, humans become more machine-like" (Harvey 2015, p. 20). Nika Mahnic connects this with the development of capitalism and commodity fetishism that involves anthropomorphizing and fetishizing inanimate market beings—corporations, for instance—on the one hand and objectifying and commodifying human beings—particularly women—on the other (Mahnic 2017). The main focus of anthropocentric humanism, therefore, is to "guard" the status of the human against objectification, de-humanization, and any form of disenchantment. Although anthropocentric humanists agree with the importance of care for (at least some) non-humans, they think it does not require using the ideas of non-anthropocentric posthumanists and is achievable in the realm of the anthropocentric humanistic approach.

Therefore, skeptics tend to remain anthropocentric humanists and oppose both the argument that robots are subjects/persons like humans and the argument that even if robots are only tools/objects, they can still replace people. From the skeptic's perspective, this leads to the disenchanting and objectification of humans. Skeptics believe that there is something special, magical, and enchanting about human beings: empathy, understanding, love, care and "human touch." These are—obviously—made possible by intrinsic features of individual human beings, but they are revealed in interactions between humans. Skeptics claim that humans still are—or at least can and should be—enchanting. They therefore oppose the disenchantment of humans and any notion of robots being enchanting.

2.5   Disenchanting Humans

Discussions examined in this chapter express the general tendency to disenchant humans, which is generally supported by intimacy robot enthusiasts and non-anthropocentric posthumanists and criticized by skeptics and anthropocentric humanists. This tendency refers both to disenchantment *with* humans and the disenchantment *of* humans.

Disenchantment *with* humans means disappointment with humans, and has been discussed here on two levels. The first level, expressed by the sociological diagnoses of human intimacy by Hochschild, Illouz, and Bauman as well as by enthusiasts of intimacy robots, concerns the disappointing character of intimate relationships with humans—these relationships are increasingly perceived, and it seems that they, at least in part, actually are, more and more difficult and painful, messy and unsatisfying, and therefore disappointing. From that perspective, intimacy robots and relationships with them are seen as a kind of relief, as a reasonable alternative, as an easier and better version of intimate life—robots provide the best things humans can provide but without the disadvantages of being with living people. The second level, expressed by non-anthropocentric posthumanists, involves the way humans treat Others and Otherness, and particularly argues that humans reduce everything to the status of a tool, property, commodity and, as a result, degrade the environment, hurt non-human life forms, and impoverish their own experience of reality. Disenchantment with humans is thus actually a disappointment with how people relate to each other (e.g., in intimate relationships), and to their surroundings (e.g., to the environment). Hence, the perspective of robots enchanting humans seem to provide an opportunity to liberate humans from disappointing relationships with other people and offers a more progressive attitude toward non-humans.

The disenchantment *of* humans involves the view of humans and their interactions as entirely calculable and therefore susceptible to engineering and duplication. These tendencies include naturalization, perceiving humans and their interactions as wholly chemical and biological processes, and economization, seeing human interactions as economical transactions. These tendencies find expression in both sociological studies of human intimacy and in enthusiasts' attitude toward intimacy robots. Simply put, if humans and their interactions are nothing more than natural processes based on calculations, there seems to be no obstacle to prevent the development of robots that will duplicate what we like in human relationships and exclude what we dislike about them. Despite the arguments of skeptics and anthropocentric humanists, however, this tendency appears to be prevalent and growing. Disenchanting humans (disenchantment with humans and the disenchantment of humans) enables robots to be enchanting for us—and to replace disenchanted humans as a source of enchantment. Yet robots are enchanting only because humans are enchanting them in the first place. The next chapter aims to understand this latter process.

REFERENCES

Allen, K. (2018, January 8). Dating Apps Closed After AI Sexbot Scam. *BBC News*. http://www.bbc.com/news/blogs-news-from-elsewhere-42609353. Accessed 23 February 2018.

Bauman, Z. (2000). *Liquid Modernity*. Cambridge, UK and Malden, MA: Polity Press.

Bauman, Z. (2003). *Liquid Love: On the Frailty of Human Bonds*. Cambridge, UK and Malden, MA: Polity Press.

Beck, U. (1992). *Risk Society: Towards a New Modernity* (M. Ritter, Trans.). London: Sage.

Beck, U., & Beck-Gernsheim, E. (1995). *The Normal Chaos of Love* (M. Ritter & J. Wiebel, Trans.). Cambridge, UK and Malden, MA: Polity Press.

Bendel, O. (2015). Surgical, Therapeutic, Nursing and Sex Robots in Machine and Information Ethics. In S. P. van Rysewyk & M. Pontier (Eds.), *Machine Medical Ethics* (pp. 17–32). Cham: Springer. https://doi.org/10.1007/978-3-319-08108-3_2.

Boellstorff, T. (2010). *Coming of Age in Second Life: An Anthropologist Explores the Virtually Human*. Princeton: Princeton University Press.

Borenstein, J., & Arkin, R. C. (2016). Robots, Ethics, and Intimacy: The Need for Scientific Research. In *Proceedings of Conference of the International Association of Computing and Philosophy (IACAP 2016)*. https://www.cc.gatech.edu/ai/robot-lab/online-publications/RobotsEthicsIntimacy-IACAP.pdf. Accessed 4 June 2017.

Borenstein, J., & Pearson, Y. (2010). Robot Caregivers: Harbingers of Expanded Freedom for All? *Ethics and Information Technology, 12*(3), 277–288. https://doi.org/10.1007/s10676-010-9236-4.

Borenstein, J., & Pearson, Y. (2011). Robot Caregivers: Ethical Issues Across the Human Lifespan. In P. Lin, K. Abney, & G. A. Bekey (Eds.), *Robot Ethics: The Ethical and Social Implications of Robotics* (pp. 251–265). Cambridge, MA and London: MIT Press.

Borenstein, J., & Pearson, Y. (2013). Companion Robots and the Emotional Development of Children. *Law, Innovation and Technology, 5*(2), 172–189. https://doi.org/10.5235/17579961.5.2.172.

Bryson, J. J. (2010). Robots Should Be Slaves. In Y. Wilks (Ed.), *Close Engagements with Artificial Companions: Key Social, Psychological, Ethical and Design Issues* (pp. 63–74). Amsterdam: John Benjamins.

Bryson, J. J., & Kime, P. P. (2011). Just an Artifact: Why Machines Are Perceived as Moral Agents. In *Proceedings of the 22nd International Joint Conference on Artificial Intelligence* (pp. 1641–1646). Menlo Park, CA: AAAI Press.

Burr-Miller, A., & Aoki, E. (2013). Becoming (Hetero) Sexual? The Hetero-Spectacle of Idollators and their Real Dolls. *Sexuality and Culture, 17*(3), 384–400. https://doi.org/10.1007/s12119-013-9187-0.

Calo, R. (2011). Robots and Privacy. In P. Lin, G. A. Bekey, & K. Abney (Eds.), *Robot Ethics: The Ethical and Social Implications of Robotics* (pp. 187–201). Cambridge, MA and London: MIT Press.

Choi, C. Q. (2008). Not Tonight, Dear, I Have to Reboot. *Scientific American, 298*(3), 94–97.

Coeckelbergh, M. (2010). Health Care, Capabilities, and AI Assistive Technologies. *Ethical Theory and Moral Practice, 13*(2), 181–190. https://doi.org/10.1007/s10677-009-9186-2.

Coeckelbergh, M. (2012). Are Emotional Robots Deceptive? *IEEE Transactions on Affective Computing, 3*(4), 388–393. https://doi.org/10.1109/t-affc.2011.29.

Coeckelbergh, M. (2015). Artificial Agents, Good Care, and Modernity. *Theoretical Medicine and Bioethics, 36*(4), 265–277. https://doi.org/10.1007/s11017-015-9331-y.

Culbertson, A. (2017, June 30). *Sex Robot Boom: Japanese Men Ditch 'Complex Women' for Sex Robots Who Don't Argue.* https://www.express.co.uk/pictures/pics/6688/Love-doll-love-hotel-Japan-pictures. Accessed 14 March 2018.

Danaher, J. (2017). Robotic Rape and Robotic Child Sexual Abuse: Should They Be Criminalised? *Criminal Law and Philosophy, 11*(1), 71–95. https://doi.org/10.1007/s11572-014-9362-x.

Danaher, J., Earp, B. D., & Sandberg, A. (2017). Should We Campaign Against Sex Robots? In J. Danaher & N. McArthur (Eds.), *Robot Sex: Social and Ethical Implications* (pp. 47–72). Cambridge, MA: MIT Press.

Decker, M. (2012). Service Robots in the Mirror of Reflective Research. *Poiesis & Praxis, 9*(3–4), 181–200. https://doi.org/10.1007/s10202-012-0111-8.

de Graaf, M. M. (2016). An Ethical Evaluation of Human-Robot Relationships. *International Journal of Social Robotics, 8*(4), 589–598. https://doi.org/10.1007/s12369-016-0368-5.

de Sio, F. S., & van Wynsberghe, A. (2016). When Should We Use Care Robots? The Nature-of-Activities Approach. *Science and Engineering Ethics, 22*(6), 1745–1760. https://doi.org/10.1007/s11948-015-9715-4.

Dépelteau, F. (2013). What Is the Direction of the "Relational Turn"? In C. Powell & F. Dépelteau (Eds.), *Conceptualizing Relational Sociology* (pp. 163–185). New York: Palgrave Macmillan.

Devlin, K. (2018). *Turned On: Science, Sex and Robots.* New York: Bloomsbury.

Enz, S., Diruf, M., Spielhagen, C., Zoll, C., & Vargas, P. A. (2011). The Social Role of Robots in the Future—Explorative Measurement of Hopes and Fears. *International Journal of Social Robotics, 3*(3), 263. https://doi.org/10.1007/s10202-012-0111-8.

Evans, D. (2010). Wanting the Impossible: The Dilemma at the Heart of Intimate Human-Robot Relationships. In Y. Wilks (Ed.), *Natural Language Processing* (pp. 75–88). Amsterdam: John Benjamins.

Fong, T., Nourbakhsh, I., & Dautenhahn, K. (2003). A Survey of Socially Interactive Robots. *Robotics and Autonomous Systems, 42*(3), 143–166.

Gerdes, A. (2016). The Issue of Moral Consideration in Robot Ethics. *Computers & Society, 45*(3), 274–279. https://doi.org/10.1145/2874239.2874278.

Giddens, A. (1991). *Modernity and Self-identity: Self and Society in the Late Modern Age.* Stanford: Stanford University Press.

Giddens, A. (1992). *The Transformation of Intimacy: Sexuality, Love and Eroticism in Modern Societies.* Stanford: Stanford University Press.

Gunkel, D. J. (2018). *Robot Rights.* Cambridge, MA and London: MIT Press.

Gutiu, S. (2012, April). *Sex Robots and Roboticization of Consent.* Paper Presented at the We Robot 2012 Conference, Coral Gables, Florida. http://robots.law.miami.edu/wp-content/2012/01/Gutiu-Roboticization_of_Consent.pdf. Accessed 22 April 2016.

Harvey, C. (2015). Sex Robots and Solipsism: Towards a Culture of Empty Contact. *Philosophy in the Contemporary World, 22*(2), 80–93. https://doi.org/10.5840/pcw201522216.

Heider, F., & Simmel, M. (1944). An Experimental Study of Apparent Behavior. *The American Journal of Psychology, 57*(2), 243–259. https://doi.org/10.2307/1416950.

Hochschild, A. R. (1983). *The Managed Heart: Commercialization of Human Feeling.* Berkeley: University of California Press.

Hochschild, A. R. (2003). *The Commercialization of Intimate Life: Notes from Home and Work.* Berkeley: University of California Press.

Hochschild, A. R. (2012). *The Outsourced Self: Intimate Life in Market Times.* New York: Picador.

Hochschild, A. R., & Ehrenreich, B. (Eds.). (2013). *Global Woman: Nannies, Maids, and Sex Workers in the New Economy.* New York: Metropolitan Books.

Holt, N. (2007). *Guys and Dolls* [TV Documentary]. A North One Television.

Illouz, E. (2007). *Cold Intimacies: The Making of Emotional Capitalism.* Cambridge, UK and Malden, MA: Polity Press.

Illouz, E. (2012). *Why Love Hurts: A Sociological Explanation*. Cambridge, UK and Malden, MA: Polity Press.

Jamieson, L. (1998). *Intimacy: Personal Relationships in Modern Societies*. Cambridge, UK: Polity Press.

Klimczyk, W. (2008). *Erotyzm ponowoczesny*. Kraków: Universitas.

Langton, R. (2009). *Sexual Solipsism: Philosophical Essays on Pornography and Objectification*. Oxford: Oxford University Press.

Laue, C. (2017). Familiar and Strange: Gender, Sex, and Love in the Uncanny Valley. *Multimodal Technologies and Interaction, 1*(1), 2. https://doi. org/10.3390/mti1010002.

Levy, D. (2007a). *Love + Sex with Robots : The Evolution of Human-Robot Relationships*. New York: Harper Perennial.

Levy, D. (2007b, April 10–14). Robot Prostitutes as Alternatives to Human Sex Workers. In *Proceedings of the IEEE-RAS International Conference on Robotics and Automation (ICRA 2007), Workshop on Roboethics, Rome, Italy*.

Levy, D. (2009). The Ethical Treatment of Artificially Conscious Robots. *International Journal of Social Robotics, 1*(3), 209–216. https://doi. org/10.1007/s12369-009-0022-6.

MacKenzie, R. (2014). Sexbots: Replacements for Sex Workers? Ethical Constraints on the Design of Sentient Beings for Utilitarian Purposes. In *Proceedings of the 2014 Workshops on Advances in Computer Entertainment Conference. ACM*. https://doi.org/10.1145/2693787.2693789.

Mahnic, N. (2017, February 3). *Sex Robot Advocacy & the Marriage of Racial Thinking & Corporate Personhood*. https://campaignagainstsexrobots. org/2017/02/03/sex-robot-advocacy-the-marriage-of-racial-thinking-cor-porate-personhood-by-nika-mahnic/. Accessed 7 December 2018.

McArthur, N. (2014). *Guiltless Pleasures of the Lonely Human Being the Moral and Legal Implications of Sex with Robots*. https://www.academia. edu/6977772/The_Moral_and_Legal_Implications_of_Sex_with_Robots. Accessed 18 October 2016.

McNair, B. (2012). *Porno? Chic!: How Pornography Changed the World and Made It a Better Place*. Abington and New York: Routledge.

Melson, G. F. (2010). Child Development Robots: Social Forces, Children's Perspectives. *Interaction Studies, 11*(2), 227–232. https://doi.org/10.1075/ is.11.2.08mel.

Misselhorn, C., Pompe, U., & Stapleton, M. (2013). Ethical Considerations Regarding the Use of Social Robots in the Fourth Age. *GeroPsych. The Journal of Gerontopsychology and Geriatric Psychiatry, 26*(2), 121–133. https://doi.org/10.1024/1662-9647/a000088.

Musiał, M. (2013). Intimacy and Modernity. Modernization of Love in the Western Culture. *Studia Europaea Gnesnensia, 7*(1), 157–168.

Musiał, M. (2015). *Intymność i jej współczesne przemiany. Studium z filozofii kultury.* Kraków: Universitas.

Musiał, M. (2018). Loving Dolls and Robots: From Freedom to Objectification, from Solipsism to Autism? In J. T. Grider & M. McDonough (Eds.), *Exploring Erotic Encounters: The Inescapable Entanglement of Tradition, Transcendence and Transgression.* Leiden: Brill Rodopi.

Nevett, J. (2018, February 1). *Paris Sex Doll Brothel Opens Despite Law BANNING Prostitution in France.* https://www.dailystar.co.uk/news/world-news/678805/paris-sex-doll-brothel-france-xdolls-prostitution-banned-law-news. Accessed 14 March 2018.

Parks, J. A. (2010). Lifting the Burden of Women's Care Work: Should Robots Replace the "Human Touch"? *Hypatia, 25*(1), 100–120. https://doi.org/10.1111/j.1527-2001.2009.01086.x.

Pearson, Y., & Borenstein, J. (2013). The Intervention of Robot Caregivers and the Cultivation of Children's Capability to Play. *Science and Engineering Ethics, 19*(1), 123–137. https://doi.org/10.1007/s11948-011-9309-8.

Pearson, Y., & Borenstein, J. (2014). Creating "Companions" for Children: The Ethics of Designing Esthetic Features for Robots. *AI & Society, 29*(1), 23–31. https://doi.org/10.1007/s00146-012-0431-1.

Pfadenhauer, M., & Dukat, C. (2015). Robot Caregiver or Robot-Supported Caregiving? *International Journal of Social Robotics, 7*(3), 393–406. https://doi.org/10.1007/s12369-015-0284-0.

Reeves, B., & Nass, C. (1996). *The Media Equation: How People Treat Computers, Television and New Media Like Real People and Places.* Cambridge, UK and New York: Cambridge University Press.

Richardson, K. (2015). *An Anthropology of Robots and AI: Annihilation Anxiety and Machines.* New York: Routledge.

Richardson, K. (2016a). Property Relations—Is Property a Person? Extending the Rights of Property Owners Through the Rights of Robotic Machines. In *Proceedings of Machine Ethics and Machine Law Workshop* (pp. 40–41). http://maschinenethik.net/wp-content/uploads/2016/11/PROCEEDINGS_MEML_2016.pdf. Accessed 2 March 2017.

Richardson, K. (2016b). Sex Robot Matters: Slavery, the Prostituted, and the Rights of Machines. *IEEE Technology and Society Magazine, 35*(2), 46–53. https://doi.org/10.1109/mts.2016.2554421.

Richardson, K. (2016c). The Asymmetrical "Relationship": Parallels Between Prostitution and the Development of Sex Robots. *Computers & Society, 45*(3), 290–293. https://doi.org/10.1145/2874239.2874281.

Richardson, K. (2016d). The Robot Intermediary: Mechanical Analogies and Autism. *Anthropology Today, 32*(5), 18–20. https://doi.org/10.1111/1467-8322.12298.

Richardson, K. (2018). *Challenging Sociality: An Anthropology of Robots, Autism, and Attachment.* New York: Palgrave MacMillan.

Richardson, K., & Gildea, F. (2017, May 5). *Sex Robots—Why We Should Be Concerned.* http://spsc.pt/index.php/2017/05/05/sex-robots-why-we-should-be-concerned/. Accessed 8 December 2018.

Robertson, J. (2010). Gendering Robots: Robo-Sexism in Japan. *Body & Society, 16*(2), 1–36. https://doi.org/10.1177/1357034x10364767.

Robins, B., Dautenhahn, K., Boekhorst, R. T., & Billard, A. (2005). Robotic Assistants in Therapy and Education of Children with Autism: Can a Small Humanoid Robot Help Encourage Social Interaction Skills? *Universal Access in the Information Society, 4*(2), 105–120. https://doi.org/10.1007/s10209-005-0116-3.

Sandel, M. J. (2012). *What Money Can't Buy: The Moral Limits of Markets.* New York: Farrar, Straus and Giroux.

Sandry, E. (2015). *Robots and Communication.* Basingstoke and New York: Palgrave Macmillan.

Scassellati, B. (2007). How Social Robots Will Help Us to Diagnose, Treat, and Understand Autism. In *Robotics Research* (pp. 552–563). Berlin and Heidelberg: Springer. https://doi.org/10.1007/978-3-540-48113-3_47.

Scheutz, M. (2011). The Inherent Dangers of Unidirectional Emotional Bonds Between Humans and Social Robots. In P. Lin, K. Abney, & G. Bekey (Eds.), *Robot Ethics: The Ethical and Social Implications of Robotics* (pp. 205–222). Cambridge, MA and London: MIT Press.

Scheutz, M., & Arnold, T. (2016). Are We Ready for Sex Robots? In *The Eleventh ACM/IEEE International Conference on Human Robot Interaction* (pp. 351–358). Piscataway, NJ: IEEE Press.

Sharkey, A. (2014). Robots and Human Dignity: A Consideration of the Effects of Robot Care on the Dignity of Older People. *Ethics and Information Technology, 16*(1), 63–75. https://doi.org/10.1007/s10676-014-9338-5.

Sharkey, A. (2016). Should We Welcome Robot Teachers? *Ethics and Information Technology, 18*(4), 283–297. https://doi.org/10.1007/s10676-016-9387-z.

Sharkey, A., & Sharkey, N. (2011). The Rights and Wrongs of Robot Care. In P. Lin, G. A. Bekey, & K. Abney (Ed.), *Robot Ethics: The Ethical and Social Implications of Robotics* (pp. 267–282). Cambridge, MA and London: MIT Press.

Sharkey, A., & Sharkey, N. (2012). Granny and the Robots: Ethical Issues in Robot Care for the Elderly. *Ethics and Information Technology, 14*(1), 27–40. https://doi.org/10.1007/s10676-010-9234-6.

Sharkey, A., & Wood, N. (2014). The Paro Seal Robot: Demeaning or Enabling. In *Proceedings of AISB 2014—50th Annual Convention of the AISB.* http://doc.gold.ac.uk/aisb50/AISB50-S17/AISB50-S17-Sharkey-Paper.pdf. Accessed 20 April 2016.

Sharkey, N., & Sharkey, A. (2006). Artificial Intelligence and Natural Magic. *Artificial Intelligence Review, 25*(1–2), 9–19. https://doi.org/10.1007/s10462-007-9048-z.

Sharkey, N., & Sharkey, A. (2010a). Living with Robots: Ethical Tradeoffs in Eldercare. In Y. Wilks (Ed.), *Natural Language Processing* (pp. 245–256). Amsterdam: John Benjamins.

Sharkey, N., & Sharkey, A. (2010b). The Crying Shame of Robot Nannies: An Ethical Appraisal. *Interaction Studies, 11*(2), 161–190. https://doi.org/10.1075/is.11.2.01sha.

Sorell, T., & Draper, H. (2014). Robot Carers, Ethics, and Older People. *Ethics and Information Technology, 16*(3), 183–195. https://doi.org/10.1007/s10676-014-9344-7.

Sparrow, R. (2002). The March of the Robot Dogs. *Ethics and Information Technology, 4*(4), 305–318. https://doi.org/10.1023/a:1021386708994.

Sparrow, R. (2016). Robots in Aged Care: A Dystopian Future? *AI & Society, 31*(4), 445–454. https://doi.org/10.1007/s00146-015-0625-4.

Sparrow, R., & Sparrow, L. (2006). In the Hands of Machines? The Future of Aged Care. *Minds and Machines, 16*(2), 141–161. https://doi.org/10.1007/s11023-006-9030-6.

Stahl, B. C., & Coeckelbergh, M. (2016). Ethics of Healthcare Robotics: Towards Responsible Research and Innovation. *Robotics and Autonomous Systems, 86*, 152–161. https://doi.org/10.1016/j.robot.2016.08.018.

Sullins, J. P. (2012). Robots, Love, and Sex: The Ethics of Building a Love Machine. *IEEE Transactions on Affective Computing, 3*(4), 398–409. https://doi.org/10.1109/t-affc.2012.31.

Truitt, E. E. (2015). *Medieval Robots: Mechanism, Magic, Nature, and Art.* Philadelphia: University of Pennsylvania Press.

Turkle, S. (2012). *Alone Together: Why We Expect More from Technology and Less from Each Other.* New York: Basic Books.

Turkle, S. (2015). *Reclaiming Conversation: The Power of Talk in a Digital Age.* New York: Penguin Press.

Vallor, S. (2011). Carebots and Caregivers: Sustaining the Ethical Ideal of Care in the Twenty-First Century. *Philosophy & Technology, 24*(3), 251. https://doi.org/10.1007/s13347-011-0015-x.

van Wynsberghe, A. (2013). Designing Robots for Care: Care Centered Value-Sensitive Design. *Science and Engineering Ethics, 19*(2), 407–433. https://doi.org/10.1007/s11948-011-9343-6.

Weeks, J. (2007). *The World We Have Won: The Remaking of Erotic and Intimate Life.* London and New York: Routledge.

Whitby, B. (2008). Sometimes It's Hard to Be a Robot: A Call for Action on the Ethics of Abusing Artificial Agents. *Interacting with Computers, 20*(3), 326–333. https://doi.org/10.1016/j.intcom.2008.02.002.

Whitby, B. (2010). Oversold, Unregulated, and Unethical: Why We Need to Respond to Robot Nannies. *Interaction Studies, 11*(2), 290–294. https://doi.org/10.1075/is.11.2.18whi.

Whitby, B. (2011). Do You Want a Robot Lover? In P. Lin, K. Abney, & G. A. Bekey (Eds.), *Robot Ethics: The Ethical and Social Implications of Robotics* (pp. 233–249). Cambridge, MA: MIT Press.

Yeoman, I., & Mars, M. (2012). Robots, Men and Sex Tourism. *Futures, 44*(4), 365–371. https://doi.org/10.1016/j.futures.2011.11.004.

YouGov. (2013). *Omnibus Poll.* http://big.assets.huffingtonpost.com/toplines-brobots.pdf. Accessed 22 July 2016.

CHAPTER 3

Humans Enchanting Robots

This chapter discusses the question of humans enchanting robots in the sense that humans think magically about them. I explain that the fact that robots are enchanting humans described in the previous chapter takes place only because humans are enchanting robots in the first place. In other words, robots are perceived as potential intimate partners only because humans think about them magically. In this sense, these two chapters are two sides of the same coin, since they describe relations between humans and robots from two different perspectives: what robots do with humans and what humans do with robots.

I begin by examining magic and magical thinking, particularly their status, their functions, their main features and the challenges of studying them. In doing so, I draw on anthropological, psychological, and philosophical accounts of magical thinking, both classic texts and more recent contributions. I also delve into reevaluations of the status of magic and magical thinking from the second part of nineteenth century to the present time and focus on shifts in its relationship with modern rational thinking. In this sense, this chapter asks the question, "How should we think about magic and magical thinking?"

Having introduced the concept, I look at the presence of magic in our interactions with and our reflections about robots. My idea of this distinction between interactions with, on the one hand, and reflections about, on the other, is based on Anna Pałubicka's (2006, 2013) distinction between two types of participation in culture. I engage

© The Author(s) 2019
M. Musiał, *Enchanting Robots*,
Social and Cultural Studies of Robots and AI,
https://doi.org/10.1007/978-3-030-12579-0_3

with two fields of inquiry here. First, the field of Human-Robot Interactions (HRI), which describes actual encounters already taking place, addressing the question "How do we think about robots (when we actually interact with them)?" Second, I discuss a specific aspect of the reflections on robot ethics developed by Mark Coeckelbergh and David Gunkel. In part because the actual interactions and experiences of robots described in HRI do not correspond to the ideas about human-robot intimacy expressed in the discussions I examined in the previous chapter, these authors attempt to rethink our thinking about humans, robots and relations, and in that sense pose the question: "How should we think about our thinking about robots?" Therefore, while the thinkers discussed in the previous chapter mainly aim to adjust the relationships with robots to our ideas of humans, robots and relationships, Gunkel and Coeckelbergh tend to adjust our thinking and ideas about robots, humans, and relations to our actual interactions with the machines. What I find common in both HRI studies and the philosophical propositions developed by Coeckelbergh and Gunkel is that they can be both interpreted as expressions of enchanting robots: both view robots as something more than mere machines. HRI scholarship shows that humans are enchanting robots in actual, spontaneous interactions with them. Gunkel and Coeckelbergh, meanwhile, suggest that it might be valuable to take these experiences and interactions seriously and to reflect and understand these robots on similar terms. I would suggest that both of these approaches can be interpreted through the lens of performing magic and magical thinking. I see similarities and analogies between the notion of magic and magical thinking on the one hand, and the ideas put forward by both HRI and Gunkel and Coeckelbergh, on the other.

In the last part of the chapter I suggest that the contemporary increase in the presence of magic and magical thinking might be a result of the disenchanting of modern rational thinking. This disenchanting includes a disenchantment *of* modern rational thinking (we no longer believe in its power) and a disenchantment *with* modern rational thinking (we are disappointed with the results of the domination of modern rational thinking). Therefore, while this chapter starts with the question "How should we think about magic and magical thinking?" it ends by addressing the question "How should we think about modern rationality and rational thinking?"

3.1 Magic and Magical Thinking

3.1.1 The Status of Magic and Magical Thinking

Wouter J. Hanegraaf, one of the most important contemporary investigators of magic, begins one of his papers by saying that magic is a wretched subject (Hanegraaff 2016). For him, magic has been—and it still is—a source of endless troubles and confusion: particularly epistemic and axiological ones. It is hard to disagree. But one must also add that magic is undoubtedly also—in part because of its wretched nature—a fascinating subject. I would now like to examine some of the issues related to magic as a subject of study. I will do that in the following order: first, clarify what I mean by magic; second, inventory the epistemic, methodological and axiological issues connected with the status ascribed to magic; and third, present a synthetic understanding of how magic works and what its functions are.

I understand magic as a mode of thinking that, in its starkest version, can be found among two groups. The first one are little children. The second group are societies that some time ago have been called "barbarians", "savages" or "primitives", and more recently "nonmoderns", "premoderns", "natives" and "indinegenous people". Since all of these labels are simplifying (at best) and abusive (at worst) I will use the term "magical societies" to address these groups of people in which the presence of magic is most starkly visible (although I do not claim, that magic is the only important feature of these groups, or the only way of thinking they are able to perform) in contrast to groups of people which were and are often labeled as "civilized", "modern", "rational" and who at least declare that they are critical towards magic and magical thinking and due to that fact the presence of magic within them is not so starkly visible (so, I do not claim there is no magic in these societies at all). I would therefore like to explore the work of anthropologists and philosophers who study the way "magical societies" think and developmental psychologists who study how children think. It is also important to notice that magical thinking is studied by psychiatrists, who point to its presence among persons with schizophrenia and obsessive-compulsive disorders. Nevertheless, I will not discuss the psychiatric account—first reason is that it would complicate the discussion that is already tricky and complex, and the second is that in psychiatry magical thinking is most often treated very one-dimensionally as a symptom of mental illness, while

I would like to deliver more nuanced and ambivalent account of its status. Therefore, I will focus on the anthropological, philosophical, and psychological ideas about the thinking of "magical societies" and children. Just for the record, I need to emphasize that by comparing children and members of "magical societies" I am taking into account only one feature that these two groups have in common: the stark presence of magical thinking. Therefore, I do not claim that "magical societies" and children are similar in general—particularly I do not claim that members of "magical societies" are "childish", or that children are "primitive".

To begin, I would like to make two introductory remarks. First, I use terms "magic" and "magical thinking" to refer to modes of thinking that are sometimes labeled in a different way. For example, Lévy-Bruhl (1975 [1949], 2015 [1910]) uses terms such as "prelogical" and "participation", and Piaget (2002 [1923], 2007 [1926]) speaks about "preoperational" and "preformal" thinking (although he also mentions magic as an aspect of both of them), while Cassirer (1955 [1946], 1960 [1925], 1972 [1944]) talks about "myth"—nevertheless I would like to use terms "magic" and "magical thinking" seem they seem the most common both in general and in contemporary studies in particular. Second, I would like to introduce a distinction that may help to clarify some matters. Although the thinking of children and "magical societies" is in many ways similar and it is no accident that anthropologists and psychologists often use the same or similar terms to describe it, there is also at least one important difference. Children's thinking, in the early stages of development, is individual matter and it is not a result of intersubjective agreement with those with whom children interact. The thinking of "magical societies", on the other hand, is for the most part intersubjectively developed and socially shared. I therefore believe it is worth distinguishing between *magical thinking*, which is an individual, supposedly innate mode of thinking, and *magic*, which is a domain of culture, a socially constructed and shared set of beliefs. Obviously, magic and magical thinking work in comparable ways; this will be discussed below. Of course, magic (a domain of culture) can be seen as an extension of magical thinking (a mode of individual reasoning), or—to put it more succinctly, magical thinking seems to be a necessary condition for magic—the latter may have not appeared without the former. What this means is that while those who are involved in magic are almost by definition also engaging in magical thinking, those who engage in magical thinking may not have anything to do with magic as a domain of culture, and may even—as we will see—be highly skeptical about its validity. To be honest, I will not refer to this distinction in most of the farther considerations much, since most of the time I will aim to understand mechanisms and patterns that refer both to magic and magical thinking, though we will use it at some point to show its fruitfulness.

The status of magic and magical thinking has changed drastically since studies of it began. Although I briefly review these transformations, what I present is by no means an exhausting account of them. Rather, I try to examine their main features and to sketch out a moderate position of my own that attempts to synthesize and balance these opposing approaches.

Studies of magic and magical thinking—understood in the sense presented above—begin in the nineteenth century with the work of two anthropologists: Edward Burnett Tylor (2010 [1971]) and James George Frazer (2002 [1890]). Both of them—as with so many intellectuals of that era—were highly influenced by Darwin's theory of evolution. This influence has—among other factors—significantly shaped the way they described the status of magical thinking. Simply put, they believed that magic represents an early stage of cultural evolution. This early character was—in their opinion—intrinsically connected with "primitivism" and the erroneous character of magic. What needs to be emphasized is that both thinkers saw magic as only *quantitatively* different from the thinking of "civilized" people—magic was simply an undeveloped version of Western modern thinking. Their accounts are sometimes labeled "intellectualistic" since they saw members of "magical societies" as persons who operate cognitively just like "civilized" people do, but whose thinking is shot through with error. What is important is that Tylor and Frazer make a concerted effort to distinguish magic from religion and science. Tylor presented "magical societies" as primitive philosophers and Frazer viewed them as primitive scientists, but both emphasized that the magic was radically fallacious and they rigidly separated it from religion and science (according to Frazer, magic, religion, and science are, in fact, consecutive evolutionary stages). According to Tylor and Frazer, therefore, the only good thing about magic was that Western society had left it behind. That is why both of them saw any remnant of magic in the culture contemporary to them as shameful and potentially dangerous regressive elements. The subsequent, functional account offered by Marcel Mauss (Mauss and Hubert 2001 [1902]) and Emile Durkheim (Durkeim and Mauss 1963 [1902]) in many ways continued the tendency to see magic as a domain of culture that is different from religion (due to its individualistic, non-social, and unorganized character) and to a lesser degree with science (due to the lack of abstract classifications). In fact, the issue of the relationship between magic, religion, and science is generally one of the most discussed aspects in the theorization of magic. Bronisław Malinowski (1954 [1948]) and Stanley Jeremiah Tambiah (1990) authored the classic accounts of it.

Other accounts of magic soon developed, as well. Ernst Cassirer (1955 [1946], 1960 [1925], 1972 [1944]) and Lucien Lévy-Bruhl (2015 [1910]) were both philosophers who became interested in what they labeled, respectively, "myth" and "primitive mentality," both of which correspond to what I call "magic" here: the particular way of thinking that seem to be starkly present in "magical societies" (however, let me emphasize again, it is neither the only way of thinking present in these societies, nor these societies are the only ones where magic and magical thinking can be observed). What is interesting about these accounts is that they—especially Lévy-Bruhl, but also Cassirer, to a lesser degree—emphasized the *qualitative* character of the differences between magic and non-magical thinking. What they showed is that many of the previous accounts of magic—mainly those of Tylor, Frazer, and their followers—simply distorted the image of magic by taking modern patterns of thought as universal, projecting them onto "magical societies" and arguing that their thinking simply represents an undeveloped, error-ridden version of a kind of thinking that was otherwise fundamentally the same. They argued that, to the contrary, magic is actually a distinct mode of thinking of its own kind. They strongly opposed the intellectualizing approach that presented "magical societies" as primitive philosophers and primitive scientists, highlighting the emotional, affective component intrinsic to magic, and also showed how modern dualisms does not apply to the syncretic character of magic (see Subsection 3, especially on Lévy-Bruhl's critique of Tylor's and Frazer's account of animism).

The further development of theories of magic in anthropology headed toward an ennoblement of magic by blurring distinctions between magical and modern thought. Claude Lévi-Strauss (1966 [1962]) is considered one of the most influential representatives of this approach. What he particularly emphasized is that magic and magical thinking can be considered neither an early stage of cultural and mental development, nor as qualitatively different from modern thinking. Magic and magical thinking, he argued, actually include most of the important elements of modern thinking. For example, he criticized the idea—shared by Mauss and Durkheim, Lévy-Bruhl and Cassirer—that "magical societies" have very limited ability to think abstractly, that they do not use highly abstract concepts, and do not build purely intellectual classifications. He believed that magic involves basically the same logic and rationality as modern thinking and is not solely interested in practical matters, but also expresses scientific curiosity. His approach—probably based on the universalism that underlay structuralism as a whole, which was founded on the claim that the structures of thought are ultimately embodied in the brain and are the same for all people—was therefore trying to show that magic and magical thinking enable their users to perform the same mental operations as modern thinking.

Finally, there is another approach that seems to have become increasingly popular in recent decades, although it was already present in Lévy-Bruhl's posthumously published "Notebooks on Primitive Mentality" (1975 [1949]). The French philosopher, after taking critiques of this previous work into account a criticism with which his previous ideas have met, decided that in fact there is no such thing as "primitive mentality." He came to the conclusion that magical thinking is typical not only of "magical societies" and their cultures, but is, in fact, a universal trait of every human mind, in addition to the kind of cognition we might call modern rational thinking. Rather than a relic of a "primitive" past, magical thinking is actually a universal mode of the human mind. This idea of two complementary modes of thinking was developed further in recent decades through a kind of a renaissance of interest in Lévy-Bruhl's thought. Anthropologists and religious studies scholars, such as Stanley J. Tambiah (1990), Wouter Hanegraaff (2003) and Susanne Greenwood (2009), have followed in Lévy-Bruhl's steps, rejecting the notion that magical thinking and magic are useless or dangerous irrational atrocities of the mind, and instead arguing that they are as important as modern rational thinking because they fulfill functions and respond to needs that modern rational thinking leaves unsolved (more on this in the next section). Hanegraaff and Greenwood—among others—also point to a persistence and, in some regards, even an increased presence of magic in Western culture (I explore this topic in detail in the next chapter).

A comparable evolution of views of magical thinking has taken place in the realm of the psychological disciplines. Sigmund Freud (1950 [1913]) in his psychoanalysis, just like Tylor and Frazer in their anthropology, considered magic to be regressive and dangerous and positing an equivalence between the thinking of "magical societies", children, and patients with psychological disorders. He saw the latter group as overwhelmed by magical thinking in result of a substantial regression. Jean Piaget (2007 [1926]), in his developmental psychology, was more careful in making comparisons, though he also saw analogies between ontogenesis (development of an individual organism) and phylogenesis (development of a group of organisms of the same species). Importantly, Piaget also considered development to be uni-directional, progressive, and substitutional. That is, he believed that the earlier stages of mental development give way to subsequent, more sophisticated stages, and that the previous stages of magical thinking (in fact, Piaget used the term "magic" in a far more narrow sense than I do it here, although the subject of his studies is exactly what is defined here as magical thinking) cease to function and disappear. Contemporary psychologists significantly modify this view (Rosengren et al. 2000). Many of them—e.g., Jaqueline Wooley or Karl Rosengren—argue that the evolution of thinking is not as linear and "progressive" as Piaget thought, that—metaphorically speaking—three steps forward are often accompanied by two—or more—steps backward. Moreover, many studies find that contemporary adults in Western societies, who consider themselves rational, actually think magically on a regular basis. Carol Nemeroff's and Paul Rozin's (1994, 2000) studies, in particular, show how the laws of magical thinking described by Frazer persist in contemporary Western societies. Eugene Subbostsky (2010, 2014) has offered probably the most complex and thorough reevaluation of magical thinking. He not only corrects Piaget's progressive and linear vision of development (as Wooley and Rosengren do) and points to the presence of magical thinking in the West (as Nemeroff and Rozin do). He also considers magical thinking to be complementary to modern rational thinking and highlights the significant functions of magical thinking that prove its usefulness (indeed, he draws on Tambiah, who—as we have seen—took similar approach in the realm of anthropology).

Moreover, it is worth mentioning that contemporary magic and magical thinking are increasingly often considered as methodological and epistemological inspiration. This refers to the discussions about neoanimism described in Chapter 5 as well as to indigenous research. Indigenous research point out the fact that thinking of "magical societies" can be considered not only as a studied object but also as a valid research perspective and may positively influence the axiological and ethical attitude of the researcher, as well as his or her epistemology and methodology,

resulting not only in an increased respect and sympathy towards the studied community and their environment, but also in a better understanding
of them. What seems particularly significant is that they enable to understand the influence which "magical societies" have had on modernity as
well as their account of modernity—a good example of that is a book significantly entitled "Why You Can't Teach United States History without
American Indians"(Sleeper-Smith et al. 2015). It helps to avoid of the
illusion of the simple vision that "we" are modern and "they" have nothing to do with neither us, nor modernity; that "we" can understand and
influence "them", but they cannot understand or influence us. Moreover,
the blurring of the distinction between "us" and "them" is not only a
result of a fact that "magical societies" have influenced us and can offer
us more than we tend to believe, but also that we have never been as
modern and disenchanted as we claim we are—this latter issue is discussed at more length in Chapter 4.

The evolution of thinking about magic and magical thinking in regard
to ascribing its status and value has thus evolved from denigrating it as
an "atavistic cultural trait" to embracing it as an "enduring quality of
the human imagination" (Brown 1997)—from a primitive and childish
relic of the past identified with regression and madness, to a useful and
important aspect of the human mind. This shift, however, is not total,
and significant voices continue to defend aspects of the evolutionary,
progressive approach. Two interesting representatives of that position are
Christopher Hallpike (1979, 2011), who offers a moderate version, and
Georg Oesterdiekhoff (2009, 2011, 2012, 2014), who presents a radical version. Both scholars build on Piaget's theory of development and
empirical cultural studies and claim that ontogenesis and phylogenesis
are comparable in regard to cognitive and moral development, or even
that the former recapitulates the latter. They contend that magical thinking is progressively replaced by a "rational", "logical" and "civilized"
mode of thinking. While Hallpike admits that the later stages of development may still involve thinking from the lower stages, Oesterdiekhoff
considers such an occurrence to be a bizarre and unlikely regression.
Both thinkers nonetheless claim that premodern magical thinking and
modern rational thinking are qualitatively different and represent consecutive stages of individual and cultural development. They thus claim that
adult members of "magical societies" are at the cognitive level of seven-
year-old children. They acknowledge, of course, that adults from "magical societies" are far more experienced and resourceful than children, but

suggest that the form in which their ideas are shaped and organized—their cognitive architecture—is the same in both cases. They claim that this conclusion is obvious and inevitable if we look at empirical studies of children's development and investigations of intellectual capabilities of members of "magical societies", and that only postmodern political correctness and excessive ideological relativism taken as an a priori dogma can obfuscate this truth.

These brief considerations suggest how fraught with difficulties studies of magic are. We can group these issues into two categories: epistemological and axiological. Epistemological issues are the problems of acquiring reliable knowledge about the thinking of members of "magical societies" and children and axiological issues are the challenges of evaluating their thinking and ascribing it a particular status, especially in comparison with modern rational thinking.

Actually, most of epistemological issues that appear in studies of magic are also at play in the study of any kind of people. When we want to understand other people, we can ask them about the things that interests us or observe what they do and how they act. Both of these approaches raise concerns. When we speak with someone we never know whether they are telling the truth (this is why we also try to check what the person actually does and whether it fits to his or her declarations), but we also cannot be sure what the person actually means, whether he or she gives words the same meanings as we do. This problem grows exponentially when we study people from a different culture than our own, when they speak a different language, or when they are little children. Moreover, the very presence of anthropologists or developmental psychologists might affect what "magical societies" or children—or, indeed, any other studied people—do. This is the truth both in reference to children who might be powerfully impacted by the unknown adult people who are asking them questions and who observe them, and to "magical societies" who inevitably may be influenced by the presence of a person with different skin color, clothes, and other features, sometimes seen for the first time by these people. At best it might result in some unnatural pretending, at worst those who are studied might become dominated by those who study and simply follow their orders in trying to fulfill what they believe is expected from them. So, when we look at how the person actually acts, we never know whether this is a spontaneous action or a "performance" for the observer. Studies of magic are thus always endangered by the risk of imputation, of the unconscious or unintended projection of our ideas onto the people we study. (I give an example of such projective imputation in Subsection 3.) Moreover, the research methods and theories that fit people from a particular culture or of a particular gender might not fit another. Examples of this can be seen in the criticism of Piaget's concept of stage development by Sudhir Kakar (1978) who has shown that it does not fit the development of Indian children, and by Carol Gilligan (1982) who has argued that Piaget's theory may refer to boys, but not to girls, who tend to develop differently—in other words, how could the results of Piaget's research be extrapolated on all children in the world, if he studied mainly children from wealthy Swiss families raised in the patriarchic Western culture? It is worth to note that this relativization is a kind of a double edged sword. For instance, Piaget is commonly criticized by psychologists claim that children are higher developed at particular ages an stages, and that Piaget has underestimated their abilities—however, it might also be the case that children studied by Piaget have actually had lower abilities than the children studied by the psychologists who criticize him.

Most of the abovementioned epistemological issues almost inevitably lead us to axiological level of problems with studying magic. It is clear and obvious that nothing—no scientific theory, in particular—is unaffected by the surrounding reality. Each thought is somehow embedded in and expresses the time and conditions in which it was developed. This is also true for accounts of magic and magical thinking. On the one hand, Randall Styers stresses that the accounts that present magic as primitive and regressive were a part of Western imperialism and moreover were a way to define modernity in opposition to it (Styers 2004). Christopher Hallpike and Georg Oesterdiekhoff, on the other hand, claim that theories of magic that describe it as complementary to modern rational thinking rather than as a stage that precedes it in the course of mental development, are ignoring empirical facts due to the political correctness and postmodern relativism. Both sides thus point to how evaluations of magic and magical thinking are embedded in broader cultural axiological tendencies. Since there are different, often contradictory tendencies in our pluralistic culture, however, they show that it is difficult to deliver a moderate and accurate account of it that is uninfluenced by ideological prejudices. Clearly, from Styers' perspective, Hallpike's and Oesterdiekhoff's theories of magic express a Western sense of superiority, while, for Hallpike and Oesterdiekhoff, Styers' approach is an example of being blinded by relativism.

It might be productive to combine some aspects of these two radical approaches. On the one hand, I do agree with Styers and other thinkers who claim that theories of magic have been an expression of Western imperialism and still can be used in this way—probably in more subtle, but no less harmful ways. I therefore agree with those who strongly oppose evaluating magic and magical thinking—as well as people who use it—as primitive, childish, inferior, erroneous, and so on. I also agree with all the approaches that see the magical thinking and modern rational thinking as complementary and equally significant parts of human mind. On the other hand, I believe that Hallpike and Oesterdiekhoff might also be right in some aspects of their developmental approach and in their claims about excessive, redundant political correctness.

I think it is hard to disagree that magic and magical thinking are earlier modes of thinking than modern rational thinking. I also believe that by saying that we do not evaluate any of these two modes of thinking in any way: by saying "earlier" we mean "preceding in time". Tylor and Frazer strongly believed that higher stages of evolution are not only later but also *better*, but we do not have to follow their hierarchical schematic. We can acknowledge the fact that some ways of thinking appeared sooner than others, yet avoid any evaluation. Of course, there is a tendency to believe that everything that is new is better—for example, that modern cars are better than bicycles. Yet, bicycles have some advantages that cars lack: they are easier to operate, they are more eco-friendly, they are easier and cheaper to buy and to repair. Of course, cars are faster, provide shelter in case of bad weather and have some other advantages over bikes. Whether bikes or cars are better depends on many factors, such as one's needs, the weather condition, distance to be taken etc. The point is that there is no way to decide—universally and once and for all—that cars are *better* than bikes (or vice versa). The same can be said about many other things: let's say movies. Can we simply say that old movies are just better or worse than contemporary ones just because they appeared earlier? By pointing this out I do not mean to go from one extreme to the other and to romanticize the past, claiming that it is better than the present—I just want to emphasize that we can simply state that some things have appeared sooner than others without evaluating them, only due to this very fact.

To complicate but at the same time to clarify this account I need to emphasize that in the previous paragraph I have compared magical thinking with *modern* rational thinking. In this sense, I do not claim that modern rationality is the only or the best rationality. Whether something rational is always better than something irrational, and whether the former is always good and the latter is always bad—as some of us may tend to believe—is a matter for a separate discussion. Nevertheless, I agree with those who argue that we cannot simply claim that members of "magical societies" and little children are utterly irrational. However, I am not sure whether it is more accurate to claim that this is because magical thinking itself is in some way rational, or because magical thinking is purely irrational, but members of "magical societies" and little children do not limit their mental activity to magical thinking and can also perform some form of rational thinking—this is also a matter for a separate investigation.

Therefore, I see magical thinking and magic as an early mode of thinking that precedes modern rational thinking. Moreover, if we agree that apart from magical thinking members of "magical societies" and little children perform also some form of rational thinking, then it seems that magical thinking is dominating in the sense that it has a wider scope and is more intensively developed in comparison to what we would label as rational thinking (on the other hand: we can argue that magic itself is in some way rational and therefore its development constitutes the development of rationality as well). However, my claims have nothing to do with saying that magic is better or worse than (modern) rationality, or that rationality is better or worse than irrationality. Actually, I believe that it is better to consider the elements of these alternatives as supplements rather than substitutes, as complementary rather than as opposite. This, in brief, is my moderate approach to the status of magical and magical thinking. One might say that my account is modern when it comes to understanding its nature, but postmodern when it comes to evaluating this nature—and this is probably the case.

As I have already mentioned, theories of magic seem to be especially vulnerable to being entangled with ideological beliefs or—to put it more broadly and neutrally—are never developed in a cultural vacuum and always express the *Zeitgeist*. From this point of view, it is worth asking what it says about our contemporary culture that we not only recognize the presence of magic and magical thinking in our culture but also recognize its positive functions. If Styers is right—and I believe he is—that modernity has been defining itself in contrast to magic, and we now increasingly consider magic to be a legitimate and useful part of our culture, it might be that we try to rethink modernity or that we no longer want to be modern. Or maybe, to echo Bruno Latour (1993), we simply recognize that "we have never been modern," and we should stop pretending that we are. In the next chapter of this book, I discuss the relationship of magic, disenchanting, re-enchanting and modernity in more detail. But next, I would like to focus on the functions of magic and magical thinking.

3.1.2   *The Functions of Magic and Magical Thinking*

Bronisław Malinowski (1978 [1935]) contributed one of the first influential examinations of the functions of magic. In his famous study, he describes Triobriand Islanders who perform magical rituals before sailing to fish in the open sea, but not when they go fishing in the nearby lagoon where the water is relatively shallow and where their boats are close to the land. With this example, Malinowski shows that magic has a larger presence when the situation is fraught with risk and uncertainty. Fishing in the lagoon is not a risky enterprise, so it does not require any magic, while fishing in the open sea is highly dangerous, so preparation for it involves magical rituals. The function of magic, according to Malinowski is therefore to provide a kind of emotional relief: to reduce feelings of anxiety and uncertainty and to provide a sense of control and prediction.

What is important is that Malinowski does not understand magic as an intellectually designed tool, but rather connects it with a universal emotional mechanism. In fact, he suggests that this mechanism is not only used by "magical societies", but also by his contemporaries and their beliefs in superstition. This suggestion has been confirmed by many scholars who have followed Malinowski's approach and who have confirmed and developed his diagnoses in their empirical studies. It has been shown that magical rituals are carried out to improve performance: the same process has been found among athletes as well as academic students preparing for exams (Damisch et al. 2010; Felson and Gmelch 1979; Gmelch 1971). An increased level of superstition has also been found among the professions where the risk is high and inevitable, such as miners, soldiers during battle, and gamblers (for a more specific description and references, see Albas and Albas 1989). Moreover, magical superstitions and rituals increase among groups of people who are faced with a serious threat, particularly in situations such as economic crises and warfare (Markle 2010). Magic thus functions as an adaptation to difficult, uncertain situations. Of course, it needs to be remembered that in many, if not most, of the abovementioned cases that pertain to contemporary Western people, the magical rituals are mainly individual procedures based on subjective emotions, experiences and interpretations, while in the case of "magical societies" they have firmly collective character.

Magic and magical thinking functions also as a way of understanding reality, as a source of explanation for why the world works the way it does, and why things happen the way they do. Since magic and magical thinking are performed by people at the early stages of human development—at both the individual level (children) and the cultural level ("magical societies")—it can be interpreted as a basic way of understanding, the most easily accessible and most rudimentary mode of grasping reality. As I have pointed out before, while this magical understanding of reality has usually been considered false and "destined" to be replaced by "rational" understanding, recently magical understandings of reality have increasingly been re-evaluated as complementary to the "rational" understanding or even as deeper and more insightful than the "rational" one. What makes the magical understanding of the world so precious, according to its proponents, is that it imbues reality with a sense of meaningfulness that contrasts with the cold and disenchanted account of reality that results from modern rational thinking, as well as a feeling of being embedded in it, of participating in it, of being an integral element of it (Greenwood 2009; Nemeroff and Rozin 2000; Subbotsky 2010).

The next function of magic may be called transgressive, in the sense of enabling something to go beyond the established order. One aspect of this function is creativity—Eugene Subbotsky (2010, 2014) emphasizes the fact that magical thinking is actually one of the most effective tools for stimulating creativity. A second aspect is—as Subbotsky calls it—the "realization of unrealized wishes." Magical thinking enables one to—at least partially—make dreams come true, in the sense that this mode of thinking enables fantasizing, but also projecting these fantasies onto reality and, in that sense, fulfilling desires that could not otherwise be fulfilled.

To summarize the functions of magic, it is important to realize that magic is increasingly seen as a supplement to modern rational thinking, rather than compensation for the lack of it. Thinkers have also increasingly emphasized that magic and magical thinking in some ways surpasses modern rational thinking by being more efficient in some matters or even by realizing functions that modern rationality is not able to carry out at all. As I stressed earlier, magic and magical thinking are more and more often considered not only increasingly present in contemporary Western culture, but also as an increasingly useful and precious part of it. Nevertheless, even though magic is not a compensation of a lack of modern rational thinking, its main functions can still be considered as a compensation of other lacks, which, in fact, reflect an adaptation to difficult conditions by providing a kind of enchantment. In the first case, the difficult situation is connected with a lack of security and a lack of control and magic compensates for these lacks by enchanting individuals to believe that they have more control than they actually have and that reality is more predictable than it actually is. As a result, magical thinking significantly improves performance. In the second case, the difficult situation is connected with a lack of meaning, a lack of understanding as well as a feeling of alienation and detachment which are compensated for by enchanting reality into a meaningful whole, in which each part is intrinsically connected with the rest. The third case is about a lack of newness and fulfillment, which is made up for by enchanting reality with new ideas and projecting unfulfilled desires onto it. I will get back to all these functions in the next chapter to show that disenchanting modernity is marked with the lack described here and that because of it, magic increases its presence and a re-enchanting of the world takes place.

3.1.3   The Functioning of Magic and Magical Thinking

One of the crucial aspects of understanding magic and magical thinking is to recognize its syncretic character. Actually, this precise term is not commonly used. Heinz Werner (2004 [1926]) is one of the few exceptions, using the term in the sense examined below. Jean Piaget (2002 [1923]) uses the term as well, but gives it a different meaning than the one presented here. The term *syncretism* refers to the fact that people who perform magic and magical thinking—at least for the time of this performance—do not make distinctions that we—contemporary Westerners—most of the time consider obvious and natural. The ways they experience the world, their worldview and the ontology that implicitly stands behind it can therefore be described in terms such as "indistinction" (Durkeim and Mauss 1963 [1902]), "undifferentiation," and "homogeneity" (Cassirer 1960 [1925]). Anna Pałubicka (1984) and

Jerzy Kmita (1984a, b), who also use the term syncretism, claim that in the cases of "magical societies" and children we should additionally see syncretism as "original" in the sense that it is not a result of a secondary mixing and connecting of what has been previously separated, but rather that it is an original and primal state of lack of distinctions—syncretism is not a consequence of connecting, but rather of a lack of disconnection. This original character of syncretism in magic and magical thinking is also emphasized by Cassirer and Lévy-Bruhl (although they put it in different terms). Syncretism among "magical societies" and children is therefore not a result of blurring distinctions, but is, rather, a consequence of the fact that no distinctions have been made yet. The situation is different with the syncretism of magic and magical thinking that takes place in the contemporary modern Western culture—it will be discussed below and in the next chapter.

Syncretism can involve several distinctions and ontological boundaries. I will discuss them next, starting from the—so it appears—most basic and fundamental ones. At the same time, I will discuss other specific features of magic and magical thinking and examine them as consequences of particular syncretisms.

The syncretism that can be considered the most basic and fundamental and, at the same time, the most surprising and counter-intuitive, is the one between the self and the world. A substantial amount of evidence provided by anthropologists and developmental psychologists confirms that those who are typically said to think magically—"magical societies" and children—do not distinguish themselves from the reality external to them. The most radical and stark example of such syncretism can be observed in children. As Piaget argues, children have no concept of there being a boundary between themselves and the rest of reality (Piaget 2007 [1926]). On the one hand, it can be said that they do not recognize their own individuality (they do not recognize themselves in the mirror, for example). On the other hand, we might argue that they do not recognize the separate character of external reality (e.g., they tend to believe that something will happen just because they think about it—Freud (1950 [1913]) labeled this latter phenomenon the "omnipotence of thought"). Piaget counters, arguing that, in fact, children *are* the world because they unaware of a boundary between themselves and the external reality—when they move their leg, they do not think it is their leg, but that it is the world. In the case of "magical societies", this syncretism and lack of distinction between the self and the rest of reality is less stark and clear, but persists in a specific form. According to significant amount of anthropological data, "magical societies" consider themselves always an integral part of a bigger whole—a family, a tribe, a totemic unity etc. They are so strongly embedded in and undifferentiated from these groups that they have difficulty grasping a scenario when something happens to themselves, but has nothing to do with the rest of the group. Christian missionaries have reported, for example, that "magical societies" could not fathom the idea of individual salvation. They believe that each of them is forever a part of his or her tribe by participating in it as a ghost or by a kind of reincarnation, making the possibility of detaching from the group after death and being judged by someone as an individual rather than as a part of a group is ridiculous and absurd.

This first syncretism—between the self and the world—leads to further syncretisms. If there is no clear distinction between the self and the object, there is also no clear distinction between the subject and the object. This, in turn, produces another group of syncretic consequences. If there is no distinction between subjects and objects, there is also no distinction between the animate and the inanimate: neither children nor "magical societies" consider themselves—or people in general, or animals, or any other beings—to be a special group of (active and animate) subjects that stand in front of and/or in opposition to (passive and inanimate) objects. Their views can be called egalitarian and non-anthropocentric since they do not privilege any group of beings—including humans—as a special entities with extraordinary ontological and/or ethical status. This is particularly visible in the case of "magical societies" who often consider non-human reality to as much a part of their society as human beings. They pray to the river to provide them fish and they ritually expiate their guilt for killing the animals they hunt. Similarly, children consider most of the artifacts they come across to be active and animate.

Another consequence of the syncretism of subject and object is undifferentiation between the internal and external. Piaget shows, for example, that children do not see their thoughts as something that exists within their minds, but rather as something that resides on the surface of the eye, in the mouth, in the air between them. The thought that refers to a particular object is a part of that object, just as the leg is a part of the chair. Finally, the syncretic account of subject and object results in lack of separation between mental and physical, spiritual and material, and so on. It is well known that children think in terms of "mind over matter" (Subbotsky 2010), since they believe that their mental states can have direct impact on the physical states (they receive a gift only because they think of it, for example, or the weather will turn rainy because they are in a sad mood). The reason they believe in the possibility of such relations is that they do not perceive the mental and the physical to be two separate ontological spheres, so they have no problem believing that one can influence the other. "Magical societies" similarly do not consider the material and the spiritual to be separate in any way. This has an important connotations for what Lévy-Bruhl labels mysticism. I will examine this below in more detail, but one example is that "magical societies" consider their dreams as at least as real as the events that take place in physical, material reality (someone who dreams about being sick, for example, might take medicine upon waking). It is not because they do not recognize that they were dreaming. In fact, they are perfectly aware of that fact. Rather, it is because they do not separate the mental and the physical the way that we do and because, for them, mental images or spiritual beings are as concrete and potentially influential as material objects and physical events.

The syncretisms I have discussed thus far—between the self and the world, the subject and the object, the animate and the inanimate, the internal and the external, the spiritual or mental and the material or physical—may be seen as the most fundamental, since they define the basic ontological features of the reality perceived from the perspective of magic and magical thinking. This reality is homogenous, undifferentiated, and does not entail the dualisms we tend to consider obvious and natural, nor does it privilege any being as a subject distinguished from mere objects. Nevertheless, the syncretism can also be found on different levels—for example, in the way the mental activities of those who perform magic and magical thinking operate. In other words, syncretism is present not only in how the world is seen, but also in the activity of the mind.

Lévy-Bruhl and Cassirer emphasized particularly strongly that affectivity is a fundamental and indispensable feature of magical existence. People who think magically do not simply know or perceive the reality—they also feel it. The word "also" in the previous sentence matters. Affectivity does not mean "magical societies" and children are emotional as opposed to intellectual, or that they feel rather than know or understand. This means that when magical thinkers experience reality, this experience is neither purely emotional, nor purely intellectual, but is syncretic, intrinsically interweaving these elements, since they haven't been distinguished and separated yet. In fact, this seemingly innate tendency toward syncretic experience, which can be observed in children, is even more stimulated in "magical societies". As Lévy-Bruhl (2015 [1910]) argues, magical thinking is embedded in collective representations, which are in fact traditions assigned to each member of a particular community during specific rituals. These rituals and collective representations are one of the conditions of what Durkheim called mechanical solidarity, and enable them to preserve their traditions and homogeneity without any fluctuations. What is characteristic of many of such rituals, which involve the transmission of content that is a part of collective representation, is that they involve strong emotions, sometimes of an extreme character. These rituals are not only connected with elevating the status of the person who participates in the ritual (for example, rites of passages from boyhood to manhood), which itself is linked with a strong emotional tension. They also stimulate emotionality, since they often involve some unusual and/or extreme circumstances or conditions such as hallucinations, torture, or fasting. In this sense, each belief that is a part of a collective representation is transmitted in a such way that it establishes the syncretic character of magical experience by connecting beliefs with strong, extreme emotions.

This is also partially an explanation for what Lévy-Bruhl calls mysticism. Mysticism is an attitude and belief shared by most if not all "magical societies" that is based in non-empirical and non-perceptual reality. These mystical beliefs are obviously a part of collective representations and are transmitted by the abovementioned rituals. As a result, "magical societies" believe without any doubt in objects, events, and powers that they cannot see, hear, or perceive. The firm character of such beliefs comes from the strong emotions that are intrinsically connected with them because of the way they became internalized. Nevertheless, it needs to be acknowledged that such magical mysticism is not the same as what we call mysticism in contemporary Western societies. When we talk about mysticism in New Age thought, Christianity, or in any other contemporary area, most of the time we mean that there is a separate, spiritual reality that it is in many ways different from the perceptible material reality. The same is true, for example, of Plato's ideas, which were embedded in a separate world. This is not the case, however, in the mysticism typical of the magic and magical thinking of "magical societies." As I have argued, the magical world is syncretic and lacking in any boundaries between the spiritual and the material. This is why the invisible, mystical reality is understood to be as real and present as what we call empirical reality. Actually, the mystical elements of reality are sometimes considered more important than what we would call the empirical aspects. In his book "How Natives Think," Lévy-Bruhl offers many examples of that phenomenon—a particularly striking one is the situation that developed after people from one tribe attacked another tribe and killed one of its members. People who were direct witnesses to the attack decided to take a revenge, but instead of simply going to the members of the tribe who they saw attacking them, they performed a ritual that directed them toward completely different tribe. Completely ignoring the empirical evidence, they attacked this other tribe to take their revenge. With children, too, it is common knowledge, not only among developmental psychologists, but also among most people who have contact with small children, that they also tend to believe in things that are not empirically perceived—invisible friends are probably most obvious example of that. The realistic status of mystical elements of the world is therefore equally obvious and undoubted for "magical societies" and children as the realistic status of empirically perceived objects is for most people in modern Western societies.

To add another level of syncretism, one Lévy-Bruhl contemplates in "How Natives Think," "magical societies" do not distinguish between natural and supernatural. This means that nothing is purely natural, or purely supernatural—each tree or river is in some sense supernatural (apart from being natural), and each spell or omen is in some sense natural (apart from being supernatural). It can therefore be said that, from our perspective, everything is mystical for "magical societies", but, from their own perspective, nothing is mystical. This is because they do not distinguish between the natural and the supernatural, the mystical and the ordinary.

The syncretic character of experience, in which an affective component is always present, is connected with the preconceptual character of magic and magical thinking. It is important to acknowledge that this preconceptual character has its more and less radical formulations. Olga Freidenberg provides an example of the former, claiming that, even in ancient Greece, abstract concepts were almost absent, and that most thoughts took the form of concrete images (Freidenberg 2004 [1978]). Lévy-Bruhl (1975 [1949]) developed a less radical approach in his notebooks, arguing that "magical societies" are able to and do form concepts, but use them differently than we do. There is nevertheless a rather common agreement, based on a substantial amount of empirical evidence, that magic and magical thinking involve a low level of abstract conceptual thinking and abstract classifications and tend to be dominated by operations with concrete images. As Emil Durkheim and Marcel Mauss showed, even when "magical societies" develop concepts, they do not use them to classify things into more general groups (species, genus). They often have names for specific kinds of trees (oaks, pines), for example, but do not have a name for trees in general, nothing that would be comparable to our word "tree," not to mention even higher levels of abstraction ("plants"). What they actually do classify most often happens in the following two ways, none of which demands a significant amount of abstract thinking. First, the classification refers to something that is present to the senses at the moment of classification and, in such a situation, most of the time, the criterion of classification is something available to the senses as well (so there is no need for a lot of abstract thinking). Second, the classification refers to thoughts that are not purely intellectual ideas, but involve a strong affective component. And, in fact, the criterion of classification has nothing to do with abstract intellectual features, but with concrete emotions connected with a specific object or event to which the idea under classification pertains. The most well-known examples of the preconceptual character of magic and magical thinking have to do with shapes and numbers. Regarding shapes, there are "magical societies" that do not have separate names for shapes, but use terms that refer to specific objects of a particular shape—for instance, they say "moon," or "plate," instead of "circle" (Luria 1976 [1974]). In this sense, they are not interested in creating an abstract name that would cover all concrete round objects, but choose the name of one of those concrete objects and use if for all of the others. As for numbers, many "magical societies" do not have words for numbers above three or four. This does not mean, however, that they cannot count. If they come into a room with dozens of people in it, for instance, they are not able to count them as we do (by giving a number to them), but they are able to remember every single person in the room. And when they come into the same room with a similar (but not the same) group of people some time later, they are able to say who is missing, who is still present, and who has joined the group. Their way of thinking is thus often based on concreteness and details instead of—as ours is—on abstractness and generality.

Another level of syncretism in magic and magical thinking has to do with its presymbolic character. Piaget, Cassirer, and Lévy-Bruhl, as well as members of the Poznań methodological school (I already mentioned Kmita and Pałubicka, but also Michał Buchowski, Wojciech J. Burszta, Andrzej P. Kowalski, and Artur Dobosz) emphasize that there is nothing symbolic about the earliest stages of magic and magical thinking, and that there is, in fact, no idea of something representing something else. As Kowalski (2001a) puts it, neither language nor thought are seen as something that has a power to represent, and, indeed, both language and though are considered to be an "ordinary" element of reality, just like stones, plants, and the like (obviously it remains closely connected with the (already examined) syncretism between subjective and objective, internal and external, spiritual/mental and material/physical). From a contemporary perspective, presymbolism can be analytically distinguished into semantic and pragmatic levels (although the "magical societies" and children who perform presymbolic operations obviously do not make such a distinction).

At the semantic level, there is no relationship of representation between the word and what the word refers to—in fact, it is even implausible to say that the word refers to anything because the word *is* the object or a part of it. It is also closely connected with the preconceptual and concrete character of magic and magical thinking—presymbolism is, in fact, another level that prevents "magical societies" and children from thinking abstractly. In his studies of languages, such as Yukaghirian, Kowalski (2001a, 2013) points out that its speakers are unable to talk about someone who is not present at the moment (i.e., there is no way to speak in the third person). One therefore always speaks about someone or something present at the moment, and if the person/object is not present at the moment it becomes present when spoken about. Verbal communication can only, therefore, be about things that are present or makes them present, since words and signs do not simply symbolize things and objects, but *are* them, or are their parts. Such studies show that symbolic thinking and the idea of symbol is a cultural invention. We are not born with it, but rather have to learn it by participating in the culture that has developed them. The story of Helen Keller is an interesting example of this kind of learning. Due to a complete loss of sight and hearing at a very young age, she was unable to learn symbolic thinking and language until the age of six, when, with the help of a teacher, Anne Sullivan, who taught her the manual alphabet, she finally discovered the importance of symbolism and language (Cassirer 1972 [1944]).

Treating words as objects or parts of objects rather than as signs that refer to objects has not disappeared completely and there are still some superstitions based on the belief that spoken words can have something to do with interacting with the object they refer to (some people, for example, do not want to use the term "cancer" when they are diagnosed in that direction since they feel it might somehow summon or activate the disease, increase the chances that it may be present). Nevertheless, while this is relatively exceptional in our thinking, it is one of the main rules for children and "magical societies". The name of an individual is not a symbolic representation, but a part of this person—that is why in many "magical societies" the name of the chief remains unknown—it could hurt the chief if his name were used in a spell or curse. A name is as concrete as a thing, because—let me emphasize this again—it *is* this thing, or part of it.

At the pragmatic level of presymbolism, speech is an action and leads to concrete consequences in physical reality. One could argue that there is nothing special about this, since, as John L. Austin has famously shown, some parts of our contemporary language are also performative actions. There are important differences here, however. In contemporary languages, speech has a result because the communication is interpreted. Speaking the words of marital vows, for example, brings a new state of affairs into being if other people understand what those spoken words mean. Similarly, the sentence "bring me the hammer please" will lead to the desired consequences only if the person who hears this speech act correctly interprets it. This is because contemporary languages are symbolic and therefore require interpretation for them to have an impact on reality. Magical speech, on the other hand, is not symbolic and—in the eyes of its users—does not require interpretation to be effective. Spells and curses are the most obvious example: even though many "magical societies" do not understand them (most of the time only the shaman does) and although these spells or curses are about objects and persons who not only are not interpreting them, but are not even hearing them and do not know they are being performed, in these cultures it is commonly believed that they directly affect the people or objects for whom they are destined. Moreover, while performatives are only one particular kind of speech act (other speech acts that are not necessarily actions can also result in transformations of reality), anthropologists hold that in magical speech, each word, sentence, and communication is an action similar to physical gesture or motion. It does not symbolically refer to or perform an action that may lead to some results if correctly interpreted; rather, it simply *is* an action that makes things happen directly and inevitably, without necessity of being understood by anyone (Bińczyk 2004; Buchowski and Burszta 1986; Malinowski 1994 [1925]).

One of the consequences of presymbolism as well as realism and concretness is one-dimensionality of the magical reality, its flatness, lack of ontological levels that are taken for granted in the realm of modern rational thinking. Since there are no abstract beings everything is concrete, since there is no fiction everything is equally real, and finally if there is no distinction between the sign and the object, then everything in the magical reality seem to have the same ontological status of concrete, real being without any ontological hierarchies (although, obviously, axiological statuses of these beings can and actually do differ a lot, yet it does not depend on the ontological status). Let me now summarize these concepts about affectivity, mysticism, preconceptualness, concreteness, presymbolism, and one-dimensionality. The affective character of magic and magical thinking comes from the syncretism of mental faculties: magical experience is never purely intellectual; it always involves strong emotions. This affectivity is one of the main conditions of mysticism, which involves the belief in non-empirical beings and is an expression of the syncretism of the natural and the supernatural, the spiritual and the material. "Magical societies" and children see no difference between what we would call the ordinary and the extraordinary. The preconceptual character of magical thinking means that "magical societies" and children do not—at least, most of the time—develop abstract concepts and even if they do, they do not build any classifications based on abstract intellectual criteria. When they develop classifications, their criteria are either based on emotional attitudes that relate to the mystical character of the classified objects or on empirical qualities of the objects that are directly present to the senses. Both of these cases highlight the concrete character of the magic and magical thinking, which avoids intellectual abstraction and focuses on specific emotional and sensual data. Finally, presymbolism means that signs and linguistic expression do not represent parts of external reality, but *are* that reality. As affectivity implies mysticism and preconceptualism implies concreteness, presymbolism implies a one-dimensionality of the magical reality. Since signs do not represent things or actions, but *are* them, or are a part of them, there is no distinction between level of meaning and level of things or actions. There is no space for interpretation and a plurality of meaning, nor is there any space for abstract concepts. The magical reality remains flat, without any ontological hierarchization. Anything you can think of, experience, or imagine is at the very same ontological level as everything else. Of course, presymbolism is a consequence of a syncretism of signs and things as well as signs and actions.

What affectivity with mysticism, preconceptualism with concreteness and presymbolism with one-dimensionality have in common (apart from their syncretic character of course) is that they are connected with a lack of abstraction, particularly with the lack of abstraction from one's affectivity in experiencing the world (inability to distance oneself from one's emotions) and lack of abstraction from concrete things (inability to distance oneself from what can be directly experienced—regardless whether it is mystical or sensual).

I would now like to point to a feature of magic and magical thinking that to a large degree can be seen as a result of already discussed affectiveness, mysticism, preconceptualism, concreteness, presymbolism, and one-dimensionality (and of syncretism connected with them as well). This feature is realism, understood as the inevitably realistic character of magic and magical thinking. By saying that this realism is inevitable, I need to emphasize that it is not the result of a conscious choice between realism and idealism or any other position. The other characteristics of magic and magical thinking make it impossible to view anything as unreal, ideal, fictional, or illusory. Magical realism is total, free of doubt, universal, and has no alternative. In fact, we can point to another level of syncretism here, one closely connected with presymbolism and one-dimensionality. There is no distinction between reality and appearance, reality and illusion, and so on. This is comparable to the lack of distinction between the natural and the supernatural. Realism, like mysticism, has no alternative. Let me explain why this is the case. The affective character of the beliefs that are a part of collective representations and a basis for magic and magical thinking are the first "legitimizations" of its realistic status. If a belief is intrinsically connected with strong emotions, it is difficult, if not impossible to have any doubt about the credibility of this belief—it is real because one feels it. This affectivity, moreover, derives from the fact that the beliefs have been transmitted during rituals and everyone in that society shares each and every belief that other member of it does—this collective character of beliefs and representations is another legitimization of its realness and unquestionableness.

Finally, presymbolism, concreteness, and one-dimensionality are probably the crucial determinants of the realistic character of magic and magic thinking. If a thought, image, sign, and, indeed, any other mental state, does not symbolize and represent anything external to it; rather, it *is* this thing (object, event, or whatever it may be), it follows that there is no space for any doubts about the ontological status of these beings. If there is no relationship of representation or even direct representation, but rather an identity of thought or sign, on the one hand, and anything else on the other, there is no question about whether one is adequate to the other or not. There is no way to say whether they fit with each other since they are not seen as separate, but as syncretically entangled without any kind of distance between them. One-dimensionality and concreteness mean that there is only one level of reality—it is simply real. There is no discussion about the status of appearance, since there is no objective or any other reality that can serve as a criterion to decide about the status of appearance. I emphasize again, therefore, that this worldview contains neither appearance nor reality, since they constitute an undifferentiated, syncretic, and homogenous whole. It can therefore be briefly said that the non-alternativity of magical realism is a result of syncretism: there is no distinction between the real and the unreal, the real and the apparent, the real and the fictional, and so on. Nor is there any distinction between internal states and the signs that represent them and the external states to which they refer. Everything is syncretic, homogenous, undifferentiated, and—by virtue of that fact—real.

To more deeply understand the rules that govern this concrete, mystical, and inevitably real world of magic, I will explore two concepts used to describe the peculiarities of magic and magic thinking: participation and metamorphosis.

Participation is a concept that was developed and explored mainly by Lévy-Bruhl, although other influential thinkers, particularly Piaget, also borrowed it. More recently, due to a kind of renaissance of interest in Lévy-Bruhl's thought, the notion of participation is becoming a subject of interest again (Greenwood 2009; Hanegraaff 2003; Tambiah 1990). The concept of participation refers to things being in constant relationships with each other, continuously connected with a set of relations—and that these relations define what these objects are. We have already seen examples of participation: a name participates with the person who owns it; doing something with a name is therefore the same as doing something with the person. This comes from the more general participation between (presymbolic) signs and things—the latter are parts of the former. The same can be said of a shadow of a particular person—it also is an integral part of him or her. Totemic beliefs are another example. Such beliefs hold that certain groups of people have strong relations with certain animals. In fact, what James George Frazer called sympathetic/homeopathic magic and contagious magic can also been seen as two kinds of participation.

The law of similarity, based on the belief that similar things can have influence on each other despite there being no contact in space between them (voodoo magic is the most obvious example of this idea) is simply based on the belief that certain things participate with each other. The same goes for the law of contagion, which holds that once things have been in contact, they remain in contact forever (if you have wounded someone with a spear, for example, putting the sharp end of the spear in a fire will make the wound worse). The laws of similarity and contagion do not exhaust the possibilities of participation, however. Mystical beliefs in almost any kind of connection—as Lévy-Bruhl and Cassirer stress—are possible, although in practice they are limited by the content of the collective representations shared in a particular society. Nevertheless, on the more general level—with both "magical societies" and children—syncretism enables the participation of everything with everything. Since there are no ontological boundaries between the realms of mind and matter, there is no reason they might have not be connected with each other and participate with each other.

As Tambiah, Greenwood and Hanegraaff suggest, participation can be seen as a way of perceiving and experiencing relations between things that is alternative to causality. But important clarifications need to be made when making this comparison. Causality is based on abstract, intellectual rules that are applied to specific cases whose rules function in practice. Participation works differently. In his early writings, Lévy-Bruhl talks about the "law of participation" and suggests that participation is a kind of a rule, an intellectually developed idea. But in his posthumously published notebooks, he explicitly turns his back on this position. He claims that there is no such thing as "law" of participation, and that participation is felt, not reflected, that it is a matter more of spontaneous emotions (based on collective representations, of course) than of intellectual considerations. Lévy-Bruhl adds, moreover, that participation is not a result of linking and connecting beings that were previously separated. Just like syncretism, participation is original in the sense that things are originally seen as participating, connected with each other. Levy-Buhl says, in fact, that the participations of a particular being define that being, but even that participations—relations with other beings—are a condition of existence. To be is to participate. As we have already acknowledged, "magical societies"' perceptions of reality, even of themselves as individuals, are always based on seeing the individual as a part of a larger entity. I mentioned earlier the Christian missionary experience that "magical societies" have problems understanding the idea of individual salvation. This is because they find it hard to imagine that the participations between an individual and his group can disappear, since the disappearance of these relations would mean the disappearance of the individual.

Another magical notion, metamorphosis, is something only briefly mentioned by most of the thinkers who deal with magic and magical thinking—Lévy-Bruhl, Cassirer, Piaget, Mauss and Durkheim, Oesterdiekhoff, and Greenwood. It is difficult to find a systematic account of it comparable to Lévy-Bruhl's description of participation. Fortunately, Andrzej P. Kowalski (1999, 2001a, b, 2013) and Artur Dobosz (2002, 2013) have recently explored this topic, developing and clarifying scattered intuitions about metamorphosis, making it a concept that contributes to the understanding of magic and magical thinking. Briefly speaking, while participation is about things being connected with each other, metamorphosis is about things becoming other things. The most famous example is from the Bororo tribe, whose members believe that they are actually becoming red parrots. This claim must be understood as literal, not metaphorical. It is not a matter of symbolism. Initially, for a people from contemporary Western culture, metamorphosis appears so ridiculous that it seems wrong to ascribe such a belief to anyone, including "magical societies".

It is worth remembering, however, that even in contemporary Western societies there are cases when metamorphosis is taken seriously. In the Catholic Church, for instance, the doctrine of transubstantiation states that during the mass the metamorphosis of the host into the body of Christ takes place literally (not symbolically as it is understood in Protestantism). There are also neo-shamanistic beliefs and other positions than can be labeled "New Age" in which metamorphosis is considered a valid experience—anthropologist Susanne Greenwood (2009) recounts her own experience of metamorphosis into a bird in one of her books. Of course, it needs to be remembered, that while in Western culture metamorphosis is highly exceptional, it is far more common among "magical societies", for whom its normality is a consequence of the characteristics of magic and magical thinking discussed above. Kowalski (2001b) and Dobosz (2002) connect it mainly with magical presymbolism. If a sign, a thought or any other mental state does not *represent* the external object, but rather *is* the object, there is no space for metaphors. The presymbolic world is one-dimensional, so nothing can be perceived as really/literally something and symbolically/metaphorically something else. Metaphorical and symbolical understandings require abstraction from the one-dimensional concreteness, and magic and magical thinking tend to avoid such abstraction. As Kowalski (2013) point outs, magic and magical thinking involve a syncretism of essence and appearance in the sense that no features of a being are more definitional or fundamental than others; all of them are equally important. None of the features ascribed to a particular being can therefore be merely metaphorical or symbolic. It is not that a person who puts on the skin of an animal and performs a specific ritual is becoming this animal only in appearance, while in essence remains a human being. Rather, the presymbolic character of magic and magical thinking does not enable to make such a distinction at all. Metamorphosis is thus a total and literal transformation into something else.

The examination of participation and metamorphosis, as well as the previous consideration, enable us to take another step in characterizing the ontology of magic and magical thinking by pointing out the relational, processual, and dynamic qualities of this reality.

Participation as a condition of existence—I participate therefore I am—in fact points to the relational character of the world seen from the perspective of magic and magical thinking. Relations of participation therefore not only define the features of a particular being, but actually also enable its existence. In this holistic world, in which everything participates (at least potentially) with everything else, there is simply no space for isolated, atomic individuals that are not entangled into the web of relations—such beings simply do not exist. We might argue, therefore, that in this mode of thinking, relations are prior to related entities in the strongest possible sense—entities do not exist without being in relationships with other entities. As for metamorphosis, it emphasizes the processual, fluid, and dynamic character of magical reality. Metamorphosis is not metaphorical fluctuation but a total and complete transformation, which means that there in nothing that stays constant after the metamorphosis; there is no essence.

This relational (due to participation) and processual (due to metamorphosis) magical reality is extremely dynamic, or "dramatic" as Cassirer (1972 [1944]) puts it and "fluid," in the words of Durkeim and Mauss (1963 [1902]). Kowalski (2001b) claims that the main units of magical ontology are events rather than objects. The search for the determinants of this dynamism leads us to another important feature of magic and magical thinking: animism. Animism is connected with the very beginning of studies of magic and magical thinking, that is, with the work of Tylor (2010 [1871]) and Frazer (2002 [1890]), but is understood differently here than in their accounts. I follow Lévy-Bruhl (2015 [1910]), who criticized Tylor and Frazer the British anthropologists for imagining that "magical societies" *perform* an animation, that something that was inanimate could—for any reason—eventually become animate. This view, according to Lévy-Bruhl, is a projection of our modern distinctions between animate and inanimate. We—modern Westerners—would be animists in Tylor's and Frazer' sense if we have decided, for example, to believe that a stone or a river were alive. But "magical societies" do not do anything like that since they do not distinguish (at least not always and not so clearly) between the animate and the inanimate. They do not *animate* anything since everything is *already and always* (at least in some sense and to some degree) alive. In other words, Tylor and Frazer do not recognize the presence of syncretism (particularly the syncretism of the animate and the inanimate) and for that reason their account of animism is a flawed projection of a modern way of thinking.

The proper understanding of animism is thus to recognize a lack of separation between the animate and the inanimate, as a result of which everything is, to some degree and in some sense, animate. What is an important consequence of animism is that all the animate objects may act according to their (or someone else's) intention. This means that magical thinking does not recognize neither objective laws based on mechanical causality nor accident. In other words, when something happens it has to have something to do with someone's intentional action. The rain cannot be explained by laws of nature or by an accidental situation—someone's intention is always behind it. This pertains even to death. There is no such thing as a natural passing, there is always someone who—most of the time by using mystical forces—has intentionally caused the death. What needs to be remembered is that it does not have to be necessarily a human being—after all, "magical societies" pray to animals that they hunt and to rivers, to not stop flowing. In a magical order, nothing happens due to unintentional causality, even river flows because it (or someone/something else) intends to make it happen.

Children also tend to think of everything as animate and only gradually expand the conditions of being animated and at the same time narrow the scope of beings that possess such a property, distinguishing them from inanimate. Children also tend to perceive the events and objects in the world in terms of intentions and functions rather than causes and properties. Piaget describes children as subscribing to artificialism, the belief that everything that is present in the world (including sun, clouds and other objects that we consider natural) was developed intentionally. They do not consider any mechanical causality to be a reasonable determinant. Piaget points out, moreover, that when children define objects, they talk about their functions, not to their physical properties—they do not say that a hammer is made of metal and wood or that it has a certain shape, but rather that it is used for pounding nails.

It can be said that "magical societies" socialize nature, although, of course, this it is not as though they decide to consider something that was formerly understood to be inanimate and/or natural to now be social. Rather, it results from syncretic understandings of subject and object, animate and inanimate, and we can add another kind of syncretism: nature and culture. In fact there is no way to distinguish the latter two realms from each other from the perspective of magic and magical thinking. Everything is natural and artificial at the same time; no being is purely natural or purely cultural. Tim Ingold describes this worldview: "there are not two worlds, of nature and society, but just one, saturated with personal powers, and embracing both humans, the animals and plants on which they depend, and the features of the landscape in which they live and move. Within this one world, humans figure not as composites of body and mind but as undivided beings, 'organism-persons', relating as such both to other humans and to non-human agencies and entities in their environment. Between these spheres of involvement there is no absolute separation, they are but contextually delimited segments of a single field" (Ingold 2000, p. 47).

In the remainder of this chapter I would like show how magic and magical thinking are present in contemporary HRI and in the ideas of Mark Coeckelbergh and David Gunkel. I aim to show how both robot users and the philosophers who are rethinking our ideas about robots are enchanting them, that is: they are acting toward them and/or thinking about them in a magical way.

3.2   Magical Thinking in Interactions with Robots

In the previous chapter, I discussed ethical and normative claims about whether and how humans should develop and relate to intimacy robots. Now, I would like to discuss studies that examine actual interactions between humans and robots that have already taken place. In exploring what lived interactions with robots look like, I am not trying to provide an answer to the debate between enthusiasts and skeptics from the previous chapter. Instead, I aim to understand these empirically observed interactions by pointing out one of the main factors that might determine their shape.

A significant number of studies conducted in the field of HRI show that humans tend to treat robots as something more than mere machines. These studies observe people of various ages (children, adults, the elderly) and robots of diverse types and levels of sophistication. They range from relatively simple service robots such as the automatic vacuum cleaner Roomba, to toys, such as the robot dog AIBO, to cutting-edge prototypes, such as Kismet or Cog developed at MIT. Below, I present a brief description of some of these studies.

AIBO is a rather unsophisticated robotic dog. Experimental studies conducted by Gail Melson, Peter H. Kahn, and their colleagues show that children tend to treat AIBO as a potential companion, ascribe it mental states and refer to it as "he" (Melson et al. 2009a, b). One interesting finding is that only some children who have treated AIBO in the abovementioned way considered it to be animate. But of the children who explicitly stated that AIBO is mechanical robot, many still treated it in a way typical of animate beings. Similar results were obtained in the study of postings about AIBO from a discussion forum where AIBO users—presumably adults—expressed attitudes toward the robot that often involved treating it as a social companion, ascribing mental sates to it, having strong emotions about it and/or considering it to be in some way animate (Friedman et al. 2003).

Sherry Turkle and her colleagues (Turkle 2006; Turkle et al. 2006b) produced similar results. They studied interactions with AIBO and My Real Baby—dolls that simulate the behavior of infants, children and the elderly. It turns out that both children and the elderly tend to develop a strong emotional attachment to both robots. They often ascribe mental states to it, nurture it, and care for it. As Turkle points out, they also attach a peculiar ontological status to the robots, saying that they are "kind of alive" or "alive enough," thus somewhere between animate and inanimate beings or being alive in their own way. Turkle suggests that the robots are blurring the boundaries between what is seen as animate and inanimate, natural and artificial, mechanical and biological. Kahn et al. (2011) point to the fact that interactions with robots are difficult to classify within our current ontological categories; they seem to go beyond them and call for new ones.

Comparable outcomes are achieved in studies of interactions between children and more sophisticated robots. Kahn, together with Hiroshi Ishiguro and colleagues, studied interactions between children and "interactive humanoid robot" Robovie (Ishiguro et al. 2001). They confirmed the tendency to ascribe mental states to the robots, and even to impute a certain moral status to them (some children felt that putting the robot into the closet when the robot was "protesting" was unfair). They also described the ontological status of Robovie in ambiguous terms. Much like with AIBO, one child said, "He's like, he's half living, half not" (Kahn et al. 2012, p. 310).

Turkle and colleagues obtained similar findings in the studies of children's interactions with Kismet and Cog. Interestingly, the children who thought the robots were animate showed strong resistance to attempts to demystify this sense of their aliveness. Not only did they ignore the robot's obvious malfunctions, rationalizing them to fit their beliefs ("he is tired" they would rationalize), but they also would not change their minds after they had been shown how Cog works and been allowed to control its movements. Some children somewhat paradoxically decided that, after they got to control Cog, they thought of him as more alive than they did when the robot was moving autonomously. As a result of this, the qualities of the object are not determined by its perceivable set of features, but rather by the way it is involved in the interaction. In Turkle's opinion, it is a shift from focusing on technical competences to focusing on relational competences (Turkle et al. 2006a).

Other published research observes similar phenomena in non-experimental, natural conditions, including studies of Roomba (an automatic vacuum cleaner) and Packbot (a military robot). The research conducted among the owners of Roomba shows that the owners surprisingly often name their devices, give them gender, develop an affectionate attitude toward them and for instance feel grateful toward their Roombas for its hard work (Sung et al. 2007). Research conducted among users of the explosives disposal robot Packbot similarly shows that soldiers feel an emotional attachment to the robot to the degree that when the specific copy they use becomes broken they disagree about replacing it with an identical one and insist on repairing their own beloved device. And when repair is impossible, they organise funerals with a honorary salvo (Carpenter 2016; Garber 2013; Garreau 2007). What is particularly interesting about the cases of Roomba and Packbot is that people perceive and/or treat them as living beings despite the fact that they resemble neither humans nor any other living organism, either in their physical appearance or in the behavior they are able to perform (none is able to simulate speech and Packbot does not even move autonomously, but is remotely controlled).

These empathetic attitudes toward Roomba and Packbot have been confirmed in experimental conditions with less sophisticated robots by Christopher Bartneck and his team (Bartneck et al. 2007a, b) as well as Cynthia Breazeal, Kate Darling and Palash Nandy (Darling 2017; Darling et al. 2015). It turns out that humans tend to be hesitant to switch off these robots, let alone to smash it with hammer or a mallet. Researchers have also observed that the time of hesitation increases with the robot's perceived level of intelligence (Bartneck et al. 2007b), the agreeableness of the robot (Bartneck et al. 2007a), and whether users were given an anthropomorphic framing (i.e., introducing the robot with a narrative including a personifying description of the robot) (Darling 2017; Darling et al. 2015).

Empathetic attitudes toward robots have also been confirmed by studies conducted by Rosenthal von der Pütten et al. (2014) as well as Suzuki et al. (2015) who used, respectively, fMRI and electroencephalography to measure subjects' level of empathy toward robots and toward human beings. They found many significant similarities when it comes to the level of empathy in both cases.

The obvious question that the results of these studies raise is how it is possible that not only children, but also adults and the elderly tend to treat robots as animate while being aware that they are dealing with a mechanical, inanimate object (of course, as I have already pointed out in the previous chapter, this phenomenon is not entirely new since humanity generally tends to treat inanimate objects and emotionally attach to them). One way to answer that question—an answer partially suggested by the last two studies described above—is to refer to biological properties of human beings, particularly to the features of the brain and to biological evolution in humans and their brains have evolved to react to particular entities in a particular way. As Turkle and Scheutz put it, surprisingly using the very same metaphorical phrase, robots "push our Darwinian buttons" (Scheutz 2011; Turkle 2010).

Acknowledging the importance of such naturalistic accounts, I would like to deliver an alternative one based on the cultural aspects of human beings. Instead of referring to the brain and its biological evolution I would like to discuss the mind and its cultural development. Despite its alternative character, however, the approach presented here is intended as complementary to the claims made by the natural sciences, not in opposition to them. Put simply, I contend that the peculiar character of human interactions with robots, described above, are a result of magical thinking performed toward these robots.

There are at least four features of the magical thinking that can be identified in the abovementioned interactions. First, animism: people tend to imagine and/or treat inanimate robots as living creatures by ascribing them mental states, moral status, or empathizing with them. What is more, at least with adults and the elderly, we can find cases of the animism that Tylor and Frazer, according to Lévy-Bruhl, mistakenly ascribed to "magical societies": animation of something previously inanimate, a clear shift in ontological status from inanimate to animate, rather than syncretic indistinction actually typical for children and "magical societies".

Second, although the robots are in some way and to some degree animated, their status is unclear: users describe them as "sort of alive," "kind of alive," "alive enough," "half living, half not." This blurring of boundaries and indifference toward beliefs we find contradictory is another aspect of magical thinking present in interactions with robots.

Third, the unclearly animated status of robots is in large degree a result of a syncretic attitude toward them in the sense that the intellectual understanding of robots seems to be intrinsically intertwined with an affective and emotional component. Users who approached the robots in a purely rational way did not see them as alive in any way and did not become attached to them. Users who perceived the robots partially through an emotional lens also saw them as partially alive, viewed them as having mental states, and empathized with them.

Fourth, partially due to the emotional element of their attitudes, users of robots were resistant to demystification and denied any proofs that the particular robot is not animate in any way—a typical approach for "classical" magic and magical thinking in the case of "magical societies" and children. "Magical societies" and children tend to completely ignore empirical evidence that contradicts their beliefs and are highly creative—to say at least—at interpreting the data in way that enables them to avoid the contradictions.

It needs to be reiterated that pointing out that magical thinking is a common element of attitudes toward technology is not an extremely new thesis. Such connections have already been shown, for instance, by Alfred Gell (1988, 1994), as well as Richard Stivers (2001)—though I am not as pessimistic about this as Stivers. More recently, Kathleen Richardson addressed the issue of animistic attitudes towards robots and labeled it "technological animism" in order "to reframe discussion of animism as a form of human consciousness that is transcultural and not unique to indigenous cosmologies" and therefore provide a "strong evidence against the enduring association between animism and the social evolutionist idea of the 'primitive'" (Richardson 2016, p. 124)—my purposes in this books are quite similar, although I prefer the terms "magic" and "magical thinking" over the term "animism". I have tried to contribute to this discussion by offering a more detailed account of magic and magical thinking, as well as an examination of the causes of the presence of magical thinking in interactions with robots and with technology in general, which I present in the next chapter.

3.3   MAGICAL THINKING IN REFLECTIONS ABOUT ROBOTS

In this section, I would like to describe the work of two philosophers whose ideas are clearly outstanding in the field of robot ethics, robophilosophy, and philosophy of technology in general: Mark Coeckelbergh and David Gunkel. What differentiates them from many other scholars—including most of those described in the previous chapter—is the perspective from which they undertake their investigations, or—in other words—the general question they ask. The thinkers described in the previous chapter, I would argue, are asking "How should we think about robots (and people)?" The HRI studies described in the previous section of this chapter, meanwhile, are trying to answer the question, "How do we—humans—think about robots in our actual interactions with them?" Gunkel and Coeckelbergh are asking something else. They recognize how we think about robots in actual, practical interactions with them and argue that it does not fit to our normative and theoretical ideas about how to think and how should we think about robots. Gunkel and

Coeckelberg therefore ask—inspired by, respectively, Derrida's deconstruction and Kant's transcendental thinking in terms of conditions of possibility (among many others)—the question "How should we think *about the way we think* about robots (and humans)?" One might say that while the thinkers discussed in the previous chapter seek to adjust our interactions with robots to how we think about them, Gunkel and Coeckelbergh try to do the opposite, to adjust our thinking about robots to how we interact with them, to rethink it in a way that enables our thinking to recognize our actual practical experience. It is worth to notice that while Coeckelbergh and Gunkel are probably the most influential thinkers who propose such a way of thinking, they are not the only ones—for example, Raya Jones (2015) shows how robots motivate us to rethink our concept of personhood and therefore asks the question: "How should we think about the way we think about subjectivity?"

My account of Gunkel's and Coeckelbergh's philosophies is admittedly incomplete, selective, and in many ways simplifying. My main purpose is to point out that some important parts of their work have strong similarities and affinities with magic and magical thinking. I argue that Coeckelbergh and Gunkel to some degree promote magical thinking due to the fact that the way of thinking they propose is in many significant regards comparable to magic. They contribute to the enchantment of robots in the sense of opening the possibility of seeing them as something more than mere machines, tools, and artifacts. I thus do not even try to do justice to Gunkel's and Coeckelbergh's ideas as a whole by attempting a thorough analysis of them. I also do not go into the differences between their approaches. Nevertheless, it is hardly controversial or surprising to highlight the similarities between their approaches since both authors recognize them themselves and moreover, others have argued that they are in agreement on some matters (see Gerdes 2016). But I also want to avoid making reductive claims that would suggest that my approach to their philosophy is a comprehensive, let alone the best account of their work. Instead, I would stress that my interpretation is just one interpretation, nothing more and nothing less. It is one possible way of looking at the "relational turn" as one way we enchant robots and as a part of broader, more general cultural transformations and tendencies. In fact, in the next chapter I argue that Gunkel's and Coeckelbergh's philosophy can be seen as part of a re-enchanting of the Western world. Again, I do not claim that this account is unquestionable or indisputable, particularly since Coeckelbergh explicitly states that it is not his intention to take part in re-enchantment or in any other romantic project—although he admits that his writings might be interpreted that way (Coeckelbergh 2017, p. 270). I am also aware that my own approach to their ideas, highlighting its similarities with magic and magical thinking, could be seen as stimulation for re-enchantment, even though that is not my intention at all. I need to make clear, moreover, that, from my perspective, the fact that thoughts or actions involve magic or magical thinking or are part of a re-enchantment of the world is cause for neither condemnation nor praise. As I try to point out in this book, particularly in this chapter and the subsequent one, magic, magical thinking and re-enchantment are highly ambiguous phenomena and my aim is to understand why they exist rather than to evaluate them.

A good place to begin is to point out what motivates both philosophers to pursue this rethinking of magical thinking, rather than remaining within the cozy and well known boundaries of modern rational thinking. Gunkel, in "Thinking Otherwise" (Gunkel 2007, pp. 41–43), has provided the best explanation of these motivations that I have found. He discusses three main motivations. One is epistemological: modern rational thinking is based on binary oppositions that excessively polarize the discourse. One side claims x (robots are people, robots are capable of feeling for people); the other side claims non-x (robots are not people, robots are not capable of feeling for people). Such polarizations, Gunkel suggests, may block productive discussion by establishing two mutually exclusive, dogmatic positions. Another reason, which has been called metaphysical, refers to the fact that these dogmatic dualisms close off opportunities created by the new phenomena. Instead of expanding our conceptual apparatus to encompass new phenomena, there is a tendency to view the novel issues reductively in order to fit them into old categories and oppositions. Gunkel and Coeckelbergh see new technologies, from the internet to robots, as an opportunity for a such reconceptualization that is most often missed. A third reason is ethical and pertains to the violent exclusions and inequalities that result from dualistic conceptualizations. According to Gunkel, when we create a dualism, one side of it always dominates the other and becomes in some sense and to some degree excluded.

What the previous paragraph makes particularly clear is that Gunkel and Coeckelbergh are highly skeptical toward dualisms and binary thinking, especially their modern expressions. They try to go beyond such binary oppositions as nature/culture and subject/object, though their main focus is on reality/appearance. They put a lot of effort into showing that there is no point in establishing that what people experience is only subjective appearances with an objective reality behind and beyond them. Gunkel, following Slavoj Žižek, holds that the expectation that there is something real beyond appearances is only a cultural construct and not a universal, indispensable need or necessity—we are not only unable to reach the things in themselves, but also not obliged to pursue that goal (Gunkel 2010). Both philosophers agree that the Platonic distinction between real and apparent can be in many cases dismissed without any loss. Indeed, they argue that we may gain by staying at the level of appearances and getting rid of the idea that there is anything real beneath, behind, or beyond. This enables us to recognize observed experiences as legitimate sources of knowledge and understanding and to see them as they are, without comparing them with a putative objective reality to decide whether or not they are worth taking into account. From this perspective, Coeckelbergh interestingly reconsiders the issue of deception in the case of intimacy robots discussed in the previous chapter (Coeckelbergh 2012a). Coeckelbergh and Gunkel therefore refrain from attempts to attain an objective reality and opt instead to take subjective appearances and concrete experiences seriously.

What follows from this phenomenological approach—as Coeckelbergh calls it—is a reevaluation of the process of obtaining knowledge. Both philosophers tend to dismiss abstract and detached theories that try to address objective reality of things themselves, in favor of a concrete and engaged practices and (inter)subjective experiences. As Coeckelbergh puts it, "'understanding' should not be taken to mean theoretical understanding but concrete experience in skilled activity" (Coeckelbergh 2012c, p. 225). What this seems to mean is that Gunkel and Coeckelbergh do not consider knowing and understanding to be a purely intellectual enterprise. When one's experience of interaction with a robot includes the tension between the knowledge that the robot is inanimate and the feeling that the robot is animate, they do not dismiss this feeling in favor of knowledge. Rather, they seem to see both knowledge and feeling as equally legitimate parts of experience, and see the experience as a basis for understanding. In fact, Coeckelbergh seems even to see emotions as more important than knowledge since he proposes a shift "from intelligent thinking to social-emotional being" (Coeckelbergh 2009, p. 217).

Another main feature of Gunkel's and Coeckelbergh's thought is often labeled "the relational turn." Generally speaking, both thinkers argue that "relations are prior to relata" and that when we want to decide on the moral status of a particular entity, we need to pay attention not to its internal objective properties, but to this entity's relations with other entities, and particularly to experiences of interactions with it. Gunkel and Cockelbergh further propose to view the world as a relational structure rather than a set of autonomous objects and to look at what happens between entities instead of focusing on their internal and objective features. Both philosophers nevertheless emphasize that relational thinking cannot become the foundation of a new objective and abstract ontology in which objects and substances are replaced with relations, but rather that what they want to focus on concrete relations that take place in actual, lived interactions, and therefore any claims made from that perspective are in fact "provisory and open to ongoing debate" (Gunkel and Cripe 2014). Therefore—let me say it once again—Coeckelbergh's and Gunkel's approach avoids reference to abstract, theoretical rules, focusing instead on concrete, practical experiences.

The phenomenological and relational approach is also very processual and dynamic. Coeckelbergh and Gunkel, as we have seen, suggest avoiding thinking in terms of stable hierarchies of abstract essences in favor of focusing on dynamic and largely unpredictable concrete relations. In fact, both philosophers suggest thinking in terms of verbs rather than in terms of nouns and adjectives. In their view, the world should not be understood as a set of entities with rigid properties, but as a dynamic set of relations and interactions. From this perspective, even values are seen not as abstract, stable beings that simply are, but as concrete and dynamic events that happen: Coeckelbergh says that "there is no truth or good (nouns); there is only *true*-ing and *good*-ing" (Coeckelbergh 2012b, p. 163). This is why Coeckelbergh uses metaphors such as "dance" or "stream of life" to describe this non-dualistic, concrete, emotional, relational and dynamic perspective of seeing reality. For them, ethics is a kind of continuous task of openness and skilled engagement that cannot be decided once and for all with a rigid set of abstract rules. As Gunkel emphasizes, ethics understood in this way shifts responsibility from the discovery of abstract rules and objective properties to decisions made by each and every individual involved in concrete relations with the other (Gunkel 2012, pp. 214–216).

This extremely brief description of Gunkel's and Coeckleberg's philosophy, without any doubt, simplifies a lot and omits even more. Yet it also highlights some of the crucial ideas of both philosophers. I would now like to show the similarities and analogies between these ideas and previously discussed magic and magical thinking, but also to point to important differences between magic in its typical context and magic in the context of thinking about the robots and thinking about thinking about robots.

One similarity is that Gunkel's and Coeckelberg's skepticism toward modern dualisms and their attempts to merge them significantly resembles the syncretic character of magic and magical thinking. Of course, we need to remember that with magic and magical thinking in their typical contexts, syncretism is original (since no particular distinctions or binary oppositions have yet been made). With the relational turn, on the other hand, the syncretism is secondary (it is a result of merging and mixing already existing binary oppositions). Another difference closely connected with this is that while the "original" magic is largely spontaneous, Coeckelbergh's and Gunkel's turn to some aspects of magic is part of a reflexive project.

Another resemblance relates to concreteness and lack of abstraction. Coeckelbergh and Gunkel avoid abstract rules and rights as well as substances and essences, in favor of concrete interactions. Moreover, by dismissing objective reality in favor of appearance, they actually approach the one-dimensional reality typical of magic, where everything is placed at the same ontological level, without any a priori hierarchies to which both philosophers formulate explicit criticism. Of course, the concreteness of typical magic and magical thinking is far more radical and its lack of abstraction is not the product of a deliberate choice, as it is in the case of relational turn.

Another resemblance has to do with a relational and dynamic vision of reality. In both cases, relations are more important than the entities connected through them and in fact these entities are defined by the relations in which they participate. There are no abstract essences that could define them. This leads to increased dynamism and thinking in terms of verbs rather than nouns and adjectives. Even the metaphors used by philosophers, such as "stream of life" resemble those used by anthropologists and psychologists to describe magical worldview ("chain of life"). The dynamism Coeckelbergh and Gunkel propose, however, is even more dynamic in one specific way: both philosophers emphasize that ethics and the ascription of moral status cannot be defined once and for all with rigid rules. In turn, primitive societies tend to preserve such rules and rights and in fact put a lot of effort into maintaining them and reject any criticism of them. Therefore, both the magical worldview and the worldview proposed by Coeckelbergh and Gunkel makes reality processual, fluid and dynamic, however contemporary philosophers make it even more dynamic, because they believe the ontological statuses of particular elements of the world can be deliberately changed, while magical societies tend to keep stability of the world in the sense that it is dynamic and fluid in the particular way, and it will remain dynamic and fluid in this very particular way: in both accounts the world is dynamic, but in the case of magical societies the rules that fund this dynamism are believed to be fixed once and for all, while Cockelbergh and Gunkel argue that any rules can be changed. Still, both approaches involve dynamic and processual worldviews.

Finally, the experiences that play a central role in the relational turn are not purely intellectual, but have an emotional component. Gunkel's and Coeckelbergh's approach—similarly to magic and magical thinking—has an affective character and treats emotions not as a source of illusions, but as a legitimate part of experience and source of knowledge. There is an important difference here, however, one that is connected with my principal doubt about Coeckelbergh's and Gunkel's philosophy. With "magical societies", emotions, feelings, and affect are largely unified since they are embedded in collective representations that are transmitted in institutionalized rituals. Emotions and experiences for them are thus generally intersubjective. If we imagine the implementation of Gunkel's and Coeckelbergh's ideas into individualistic Western societies it is hard to say what the intersubjectivity of emotions, experiences, and general worldview would be based on. It would seem that instead of living in one world, individuals may find themselves encapsulated in their own solipsistic or even autistic worlds (see my discussion of solipsism in the previous chapter). If emotions and personal experience is the main source of knowledge about the world, and there is no uniformization of the emotions and experiences, how could individuals understand themselves as living in the same world? One could say that their experience does not matter as much as the fact that, objectively, they actually live in the same world and this is enough safeguard for inter-subjectivity. What Gunkel and Coeckelbergh argue, however, is that relations precede relata, ethics precede ontology, experience creates reality. Assuming that intersubjectivity is not given but is developed, there is a risk that an implementation of the relational turn might lead—contrary to what Coeckelbergh claims about the reduction of alienation—to an increase in detachment, especially detachment from other people and their subjective realities.

Nevertheless, the question of thinking otherwise seems to be one of the most influential parts of the contemporary philosophy of technology that can be included in a broader standpoint that I have called anti-an-thropocentric posthumanism in the previous chapter. I believe that to pursue further in this direction it might be worth looking not only at premodern societies and their magic and magical thinking, but perhaps also at contemporary non-Western societies that have gone through a different experience of modernization, such as the Japanese, whose particular attitude toward robots and technology Jennifer Robertson (2017) has described thoroughly. Others worth bringing in include modern theorists such as Gregory Bateson, whose work—according to Morris Berman (1981) and Susan Greenwood (2009)—remains unexplored as an inspiration for the transformation of modern thinking (Bateson 1999 [1972]).

3.4   Disenchanting Modern Rationality

This chapter examined magic and magical thinking and has shown its presence in interacting with and reflecting about robots, as well as a reevaluation of this presence and of magic and magical thinking itself. Without any doubt, this shift in our thinking about magic and magical thinking results from and has an impact on the way we think about the thinking that is traditionally considered the opposite of magic and magical thinking: modern rationality and modern rational thinking. I do not want to go into the various definitions of rationality and rational thinking, not mentioning the issue concerning rationality of magic and magical thinking intensively debated by cultural and social anthropologists. Let me simply remark that what I mean here is thinking based on logic and/or empirical evidence that is considered an intrinsic element of the modern idea of progress, particularly with scientific and technological progress. Rationality is a mode of thinking that modernity enthroned as progressive and useful, as opposed to magical thinking, seen as primitive and useless. Moreover, as I have pointed out a few times, I emphasize that I assume there can be other rationalities than the modern one and

I do not think that everything that does not fit to modern rationality—including magical thinking—needs to be considered as irrational. This chapter can be seen as a diagnosis of the tendency to disenchant modern rationality, which contains two levels: the disenchantment *of* modern rationality and its high status, and disenchantment *with* modern rationality and its actual results. As has been shown, particularly in the first section, modern rational thinking is not so universally useful and it is often thought that magical thinking is far more effective in performing certain functions and therefore the efficiency and usefulness of modern rationality is quite narrow and far from universal. We see that modern rationality is not always—or, in fact, quite rarely—the best solution for our problems and the tasks that stand in front of us. We can thus observe the disenchantment of modern rationality as the most universally efficient mode of thinking. Moreover, we point out the negative features and consequences of modern rational thinking. We increasingly see it as leading to the exclusion and objectification of Others, as a part of the domination of humans over other beings and the environment and their increasingly degrading—but also rational and efficient—means of exploitation. In this sense, we are disenchanted with modern rational thinking and its cumulative result: we no longer believe that modern rational thinking always makes the world a better place.

Obviously, the disenchantment of and with modern rationality is connected with re-evaluations of notions such as progress, humanism, anthropocentrism, and logocentrism—core elements of modernity and modern thinking. In the next chapter, I would like to examine this broader perspective and discuss how we rethink and re-evaluate our thinking about modernity, by pointing out the processes of disenchanting modernity, as well as to the presence of disenchantment and re-enchantment in the modern world.

References

Albas, D., & Albas, C. (1989). Modern Magic: The Case of Examinations. *Sociological Quarterly, 30*, 603–613.

Bartneck, C., van der Hoek, M., Mubin, O., & Al Mahmud, A. (2007a). "Daisy, Daisy, Give Me Your Answer Do!": Switching Off a Robot. In *Proceedings of the 2nd ACM/IEEE International Conference on Human-Robot Interaction* (pp. 217–222). Washington, DC: ACM Press. https://doi.org/10.1145/1228716.1228746.

Bartneck, C., Verbunt, M., Mubin, O., & Al Mahmud, A. (2007b). To Kill a Mockingbird Robot. In *Proceedings of the 2nd ACM/IEEE International Conference on Human-Robot Interaction* (pp. 81–87). Washington, DC: ACM Press. https://doi.org/10.1145/1228716.1228728.

Bateson, G. (1999 [1972]). *Steps to an Ecology of Mind*. Chicago and London: University of Chicago Press.

Berman, M. (1981). *The Reenchantment of the World*. Ithaca, NY: Cornell University Press.

Bińczyk, E. (2004). Mowa magiczna a referencja. In A. Pałubicka & A. Dobosz (Eds.), *Umysł i kultura* (pp. 175–186). Bydgoszcz: Oficyna Wydawnicza Epigram.

Brown, M. E. (1997). Thinkinq About Maqic. In S. D. Glazier (Ed.), *Anthropology of Religion: A Handbook*. Westport: Greenwood Press.

Buchowski, M., & Burszta, W. J. (1986). Status języka w świadomości magicznej. *Etnografia polska, 30*(1), 31–42.

Carpenter, J. (2016). *Culture and Human-Robot Interaction in Militarized Spaces: A War Story*. Surrey, Burlington: Ashgate.

Cassirer, E. (1955 [1946]). *The Myth of the State*. Garden City, NY: Doubleday.

Cassirer, E. (1960 [1925]). *The Philosophy of Symbolic Form. Volume 2, Mythical Thought* (C. W. Hendel, Trans.). New Haven: Yale University Press.

Cassirer, E. (1972 [1944]). *An Essay on Man: An Introduction to a Philosophy of Human Culture*. New Haven: Yale University Press.

Coeckelbergh, M. (2009). Personal Robots, Appearance, and Human Good: A Methodological Reflection on Roboethics. *International Journal of Social Robotics, 1*(3), 217–221. https://doi.org/10.1007/s12369-009-0026-2.

Coeckelbergh, M. (2012a). Are Emotional Robots Deceptive? *IEEE Transactions on Affective Computing, 3*(4), 388–393. https://doi.org/10.1109/t-affc.2011.29.

Coeckelbergh, M. (2012b). *Growing Moral Relations: Critique of Moral Status Ascription*. Basingstoke and New York: Palgrave Macmillan.

Coeckelbergh, M. (2012c). Technology as Skill and Activity: Revisiting the Problem of Alienation. *Techné: Research in Philosophy and Technology, 16*(3), 208–230. https://doi.org/10.5840/techne201216315.

Coeckelbergh, M. (2017). *New Romantic Cyborgs: Romanticism, Information Technology, and the End of the Machine*. Cambridge, MA and London: MIT Press.

Damisch, L., Stoberock, B., & Mussweiler, T. (2010). Keep Your Fingers Crossed! How Superstition Improves Performance. *Psychological Science, 21*(7), 1014–1020. https://doi.org/10.1177/0956797610372631.

Darling, K. (2017). "Who's Johnny?" Anthropomorphic Framing in Human-Robot Interaction, Integration, and Policy. In K. Abney, P. Lin, & G. A. Bekey (Eds.), *Robot Ethics 2.0*. Oxford: Oxford University Press.

Darling, K., Nandy, P., & Breazeal, C. (2015). Empathic Concern and the Effect of Stories in Human-Robot Interaction. In *Proceeding of the IEEE International Workshop on Robot and Human Communication (ROMAN)* (pp. 770–775). https://doi.org/10.1109/roman.2015.7333675.

Dobosz, A. (2002). *Tożsamość metamorficzna a komunikacja językowa.* Poznań: Wydawnictwo Naukowe Instytutu Filozofii UAM.

Dobosz, A. (2013). *Myślenie magiczno-mityczne a schizofrenia.* Bydgoszcz: Oficyna Wydawnicza Epigram.

Durkeim, E., & Mauss, M. (1963 [1902]). *Primitive Classification* (R. Needham, Trans.). Chicago: University of Chicago Press.

Felson, R. B., & Gmelch, G. (1979). Uncertainty and the Use of Magic. *Current Anthropology, 20*(3), 587–589.

Frazer, J. G. (2002 [1890]). *The Golden Bough.* Mineola, NY: Dover Publications.

Freidenberg, O. (2004 [1978]). *Image and Concept: Mythopoetic Roots of Literature.* Oxon: Routledge.

Freud, S. (1950 [1913]). *Totem and Taboo: Some Points of Agreement Between the Mental Lives of Savages and Neurotics* (J. Strachey, Trans.). New York: W.W. Norton.

Friedman, B., Kahn P. H. Jr., & Hagman, J. (2003). Hardware Companions? What Online AIBO Discussion Forums Reveal About the Human-Robotic Relationship. In *Proceedings of the SIGCHI Conference on Human Factors in Computing Systems* (pp. 273–280). New York, NY: ACM.

Garber, M. (2013, September 20). Funerals for Fallen Robots: New Research Explores the Deep Bonds That Can Develop Between Soldiers and the Machines that Help Keep Them Alive. *The Atlantic.* https://www.theatlantic.com/technology/archive/2013/09/funerals-for-fallen-robots/279861/. Accessed 25 July 2015.

Garreau, J. (2007, May 6). Bots on the Ground. *The Washington Post.* http://www.washingtonpost.com/wp-dyn/content/article/2007/05/05/AR2007050501009.html. Accessed 25 July 2015.

Gell, A. (1988). Technology and Magic. *Anthropology Today, 4*(2), 6–9. https://doi.org/10.2307/3033230.

Gell, A. (1994). The Technology of Enchantment and the Enchantment of Technology. In J. Coote (Ed.), *Anthropology, Art, and Aesthetics.* Oxford: Clarendon Press.

Gerdes, A. (2016). The Issue of Moral Consideration in Robot Ethics. *Computers and Society, 45*(3), 274–279. https://doi.org/10.1145/2874239.2874278.

Gilligan, C. (1982). *In a Different Voice. Psychological Theory and Women's Development.* Cambridge, MA: Harvard University Press.

Gmelch, G. (1971). Baseball Magic. *Transaction, 8*(8), 39–41.

Greenwood, S. (2009). *The Anthropology of Magic*. Oxford and New York: Bloomsbury Academic.

Gunkel, D. J. (2007). *Thinking Otherwise: Philosophy, Communication, Technology*. West Lafayette, IN: Purdue University Press.

Gunkel, D. J. (2010). The Real Problem: Avatars, Metaphysics and Online Social Interaction. *New Media and Society, 12*(1), 127–141. https://doi.org/10.1177/1461444809341443.

Gunkel, D. J. (2012). *The Machine Question: Critical Perspectives on AI, Robots, and Ethics*. Cambridge, MA: MIT Press.

Gunkel, D. J., & Cripe, B. (2014). Apocalypse Not, or How I Learned to Stop Worrying and Love the Machine. *Kritikos, 11*.

Hallpike, C. R. (1979). *The Foundations of Primitive Thought*. Oxford and New York: Oxford University Press.

Hallpike, C. R. (2011). *On Primitive Society: And Other Forbidden Topics*. Bloomington, IN: AuthorHouse.

Hanegraaff, W. J. (2003). How Magic Survived the Disenchantment of the World. *Religion, 33*(4), 357–380. https://doi.org/10.1016/s0048-721x(03)00053-8.

Hanegraaff, W. J. (2016). Magic. In G. Magee (Ed.), *The Cambridge Handbook of Western Mysticism and Esotericism* (pp. 393–404). Cambridge: Cambridge University Press.

Ingold, T. (2000). *The Perception of the Environment: Essays on Livelihood, Dwelling and Skill*. London and New York: Routledge.

Ishiguro, H., Ono, T., Imai, M., Maeda, T., Kanda, T., & Nakatsu, R. (2001). Robovie: An Interactive Humanoid Robot. *Industrial Robot: The International Journal of Robotics Research and Application, 28*(6), 498–504. https://doi.org/10.1108/01439910110410051.

Jones, R. A. (2015). *Personhood and Social Robotics: A Psychological Consideration*. Abingdon, New York: Routledge.

Kahn, P. H., Reichert, A. L., Gary, H. E., Kanda, T., Ishiguro, H., Shen, S., & Gill, B. (2011). The New Ontological Category Hypothesis in Human-Robot Interaction. In *2011 6th ACM/IEEE International Conference on Human-Robot Interaction (HRI)* (pp. 159–160). IEEE. https://doi.org/10.1145/1957656.1957710.

Kahn, P. H. Jr., Kanda, T., Ishiguro, H., Freier, N. G., Severson, R. L., Gill, B. T., & Shen, S. (2012). "Robovie, You'll Have to Go into the Closet Now": Children's Social and Moral Relationships with a Humanoid Robot. *Developmental Psychology, 48*(2), 303. https://doi.org/10.1037/a0027033.

Kakar, S. (1978). *Inner World: A Psycho-Analytic Study of Childhood and Society in India: Psychoanalytic Study of Childhood and Society in India*. Delhi: Oxford University Press.

Kmita, J. (1984a). Magiczne źródło kultury. *Odra, 2*, 24–30.

Kmita, J. (1984b). Mowa magiczna—język—literatura. *Odra, 12*, 32–37.

Kowalski, A. P. (1999). *Symbol w kulturze archaicznej.* Poznań: Wydawnictwo Naukowe Instytutu Filozofii UAM.

Kowalski, A. P. (2001a). *Myślenie przedfilozoficzne: studia z filozofii kultury i historii idei.* Poznań: Wydawnictwo Fundacji Humaniora.

Kowalski, A. P. (2001b). O wyobrażeniu metamorfozy i o doświadczeniu magicznym w kulturze wczesnotradycyjnej. In J. Kmita (Ed.), *Czy metamorfoza magiczna rekompensuje brak symbolu?* (pp. 19–93). Poznań: Wydawnictwo Naukowe Instytutu Filozofii UAM.

Kowalski, A. P. (2013). *Mit a piękno: z badań nad pochodzeniem sztuki.* Bydgoszcz: Oficyna Wydawnicza Epigram.

Latour, B. (1993). *We Have Never Been Modern* (C. Porter, Trans.). Cambridge, MA: Harvard University Press.

Lévi-Strauss, C. (1966 [1962]). *The Savage Mind.* Chicago: University of Chicago Press.

Lévy-Bruhl, L. (1975 [1949]). *The Notebooks on Primitive Mentality* (M. Leenhardt, Ed., P. Riviere, Trans.). New York: Harper & Row.

Lévy-Bruhl, L. (2015 [1910]). *How Natives Think* (L. A. Clare, Trans.). Eastford, CT: Martino Fine Books.

Luria, A. R. (1976 [1974]) *Cognitive Development: Its Cultural and Social Foundations* (M. Lopez-Morillas & L. Solotaroff, Trans.). Cambridge, MA and London: Harvard University Press.

Malinowski, B. (1954 [1948]). *Magic, Science and Religion and Other Essays* (R. Redfield, Ed.). Garden City, NY: Doubleday.

Malinowski, B. (1978 [1935]). *Coral Gardens and Their Magic: A Study of the Methods of Tilling the Soil and of Agricultural Rites in the Trobriand Islands: Two Volumes Bound As One.* New York: Dover Publications.

Malinowski, B. (1994 [1923]). The Problem of Meaning in Primitive Languages. In C. K. Ogden & I. A. Richards (Eds.), *The Meaning of Meaning* (pp. 296–336). San Diego, New York and London: HBJ Book.

Markle, D. T. (2010). The Magic That Binds Us: Magical Thinking and Inclusive Fitness. *Journal of Social, Evolutionary, and Cultural Psychology, 4*(1), 18–33. https://doi.org/10.1037/h0099304.

Mauss, M., & Hubert, H. (2001 [1902]). *A General Theory of Magic.* London and New York: Routledge.

Melson, G. F., Kahn, P. H. Jr., Beck, A., & Friedman, B. (2009a). Robotic Pets in Human Lives: Implications for the Human–Animal Bond and for Human Relationships with Personified Technologies. *Journal of Social Issues, 65*(3), 545–567.

Melson, G. F., Kahn, P. H. Jr., Beck, A., Friedman, B., Roberts, T., Garrett, E., & Gill, B. T. (2009b). Children's Behavior Toward and Understanding of Robotic and Living Dogs. *Journal of Applied Developmental Psychology, 30*(2), 92–102. https://doi.org/10.1016/j.appdev.2008.10.011.

Nemeroff, C., & Rozin, P. (1994). The Contagion Concept in Adult Thinking in the United States: Transmission of Germs and of Interpersonal Influence. *Ethos, 22*(2), 158–186. https://doi.org/10.1525/eth.1994.22.2.02a00020.

Nemeroff, C., & Rozin, P. (2000). The Making of the Magical Mind: The Nature and Function of Sympathetic Magical Thinking. In K. S. Rosengren, C. N. Johnson, & P. L. Harris (Eds.), *Imagining the Impossible: Magical, Scientific and Religious Thinking in Children* (pp. 1–34). Cambridge: Cambridge University Press.

Oesterdiekhoff, G. W. (2009). *Mental Growth of Humankind in History.* Norderstedt: BoD–Books on Demand.

Oesterdiekhoff, G. W. (2011). *The Steps of Man Towards Civilization.* Norderstedt: BoD—Books on Demand.

Oesterdiekhoff, G. W. (2012). Was Pre-Modern Man a Child? The Quintessence of the Psychometric and Developmental Approaches. *Intelligence, 40*(5), 470–478.

Oesterdiekhoff, G. W. (2014). The Role of Developmental Psychology to Understanding History, Culture and Social Change. *Journal of Social Sciences, 10*(4), 185–195. https://doi.org/10.3844/jssp.2014.185.195.

Pałubicka, A. (1984). *Przedteoretyczne postaci historyzmu.* Warszawa: Państwowe Wydawnictwo Naukowe.

Pałubicka, A. (2006). *Myślenie w perspektywie poręczności a pojęciowa konstrukcja świata.* Bydgoszcz: Oficyna Wydawnicza Epigram.

Pałubicka, A. (2013). *Gramatyka kultury europejskiej.* Bydgoszcz: Oficyna Wydawnicza Epigram.

Piaget, J. (2002 [1923]). *The Language and Thought of the Child* (M. Gabain & R. Gabain, Trans.). London and New York: Routledge.

Piaget, J. (2007 [1926]). *The Child's Conception of the World: A 20th-Century Classic of Child Psychology* (J. Tomlinson & A. Tomlinson, Trans.). Lanham, MD: Rowman & Littlefield.

Richardson, K. (2016). Technological Animism: The Uncanny Personhood of Humanoid Machines. *Social Analysis, 60*(1), 110–128. https://doi.org/10.3167/sa.2016.600108.

Robertson, J. (2017). *Robo sapiens japanicus: Robots, Gender, Family, and the Japanese Nation.* Berkeley: University of California Press.

Rosengren, K. S., Johnson, C. N., & Harris, P. L. (Eds.). (2000). *Imagining the Impossible: Magical, Scientific, and Religious Thinking in Children.* Cambridge, UK and New York: Cambridge University Press.

Rosenthal-von der Pütten, A. M., Schulte, F. P., Eimler, S. C., Sobieraj, S., Hoffmann, L., Maderwald, S., et al. (2014). Investigations on Empathy Towards Humans and Robots Using fMRI. *Computers in Human Behavior, 33*, 201–212. https://doi.org/10.1016/j.chb.2014.01.004.

Scheutz, M. (2011). The Inherent Dangers of Unidirectional Emotional Bonds between Humans and Social Robots. In P. Lin, K. Abney, & G. Bekey (Eds.),

Robot Ethics: The Ethical and Social Implications of Robotics (pp. 205–222). Cambridge and London: MIT Press.

Sleeper-Smith, S., Barr, J., O'Brien, J. M., Shoemaker, N., & Stevens, S. M. (Eds.). (2015). *Why You Can't Teach United States History Without American Indians*. Chapel Hill: University of North Carolina Press.

Stivers, R. (2001). *Technology as Magic: The Triumph of the Irrational*. New York and London: Bloomsbury Academic.

Styers, R. (2004). *Making Magic: Religion, Magic, and Science in the Modern World*. Oxford and New York: Oxford University Press.

Subbotsky, E. (2010). *Magic and the Mind: Mechanisms, Functions, and Development of Magical Thinking and Behavior*. Oxford and New York: Oxford University Press.

Subbotsky, E. (2014). The Belief in Magic in the Age of Science. *SAGE Open, 4*(1). https://doi.org/10.1177/2158244014521433.

Sung, J.-Y., Guo, L., Grinter, R. E., & Christensen, H. I. (2007). "My Roomba Is Rambo": Intimate Home Appliances. In J. Krumm, G. D. Abowd, A. Seneviratne, & T. Strang (Eds.), *Proceedings of UbiComp 2007: Ubiquitous Computing* (pp. 145–162). Berlin and Heidelberg: Springer. https://doi.org/10.1007/978-3-540-74853-3_9.

Suzuki, Y., Galli, L., Ikeda, A., Itakura, S., & Kitazaki, M. (2015). Measuring Empathy for Human and Robot Hand Pain Using Electroencephalography. *Scientific Reports, 5*, 15924. https://doi.org/10.1038/srep15924.

Tambiah, S. J. (1990). *Magic, Science and Religion and the Scope of Rationality*. Cambridge, UK and New York: Cambridge University Press.

Turkle, S. (2006). *A Nascent Robotics Culture: New Complicities for Companionship* (AAAI Technical Report Series).

Turkle, S. (2010). In Good Company? On the Threshold of Robotic Companions. In Y. Wilks (Ed.), *Close Engagements with Artificial Companions: Key Social, Psychological, Ethical and Design Issues* (pp. 3–10). Amsterdam: John Benjamins.

Turkle, S., Breazeal, C., Dasté, O., & Scassellati, B. (2006a). Encounters with Kismet and Cog: Children Respond to Relational Artifacts. *Digital Media: Transformations in Human Communication*, 1–20.

Turkle, S., Taggart, W., Kidd, C. D., & Dasté, O. (2006b). Relational Artifacts with Children and Elders: The Complexities of Cybercompanionship. *Connection Science, 18*(4), 347–361. https://doi.org/10.1080/09540090600868912.

Tylor, E. (2010 [1871]). *Primitive Culture: Researches into the Development of Mythology, Philosophy, Religion, Art, and Custom*. Cambridge, UK: Cambridge University Press.

Werner, H. (2004 [1926]). *Comparative Psychology of Mental Development*. Clinton Corners, NY: Percheron Press.

CHAPTER 4

Disenchanting and Re-enchanting in Modernity

The previous chapter identified the presence of magical thinking both at the level of engaged, spontaneous interactions with robots and at the level of philosophical, theoretical reflections about them. I have also described how the status of magical thinking has been recently reevaluated from a primitive and childish relic of the past to a valuable and ever-present mode of thinking complementary to what can be called modern rational thinking. In this sense the previous chapter has been an answer to the questions: "How should we think about the way we think about robots?" and "How should we think about the way we think about our (magical and rational) thinking?" This chapter in some ways follows the previous one by expanding the discussion about the presence and status of magic from the field of interactions with and reflections about robots to the more general level of contemporary Western culture. More specifically, it examines the issue of our ideas of modernity and the processes of disenchantment and re-enchantment as main parts of it. In particular, it makes the claim that modernity can be understood in terms of a coexistence of disenchantment and re-enchantment, and that these are not only oppositional, but also in many ways complementary.

This chapter thus focuses on the status of magic in modernity, and on the status of modernity in general, by examining analyses that point to the disenchantment of the world (or the lack of it) and/or to the

© The Author(s) 2019   115
M. Musiał, *Enchanting Robots*,
Social and Cultural Studies of Robots and AI,
https://doi.org/10.1007/978-3-030-12579-0_4

re-enchantment of the world. After pointing out the complementary character of most of this work—that is, observing that both disenchantment and re-enchantment are taking place—the chapter discusses the causes of these processes as a way of explaining the abovementioned situation. The two previous chapters discussed the issue of enchantments taking place between humans and robots; this chapter discusses the enchantments (as well as the dis- and re-enchantments) between humans and "the rest of the world," to answer the question, "How should we think about the way we think about disenchantment, re-enchantment, and modernity?" While the chapter "Robots Enchanting Humans" explores the processes of disenchantment of and with human beings, and the chapter "Humans Enchanting Robots" investigates the processes of disenchantment of and with modern rational thinking, this chapter focuses on the disenchantment of and with modernity.

The chapter is structured as follows: I first examine diagnoses that discuss re-enchantment and disenchantment and argue that the two processes should be understood not as opposing, but as complementary. I contend that this is true at the local and global levels, as well as at the level of interactions with and reflections about the world. Finally, I point to the causes of re-enchantment by referring to the functions of magic that compensate for some lacks in contemporary culture, caused by extensive disenchantment.

4.1 Disenchantment and Its Limits

The metaphor of the "disenchantment of the world" popularized by Max Weber (though he borrowed it from Friedrich Schiller) has had an astonishing career in the discourse of the humanities and social sciences. Interestingly, Weber himself only used the metaphor a few times (for the complete catalogue, see Sherry 2009), and did not offer any extended elaboration (Hanegraaff 2003, p. 358). Weber's most popular and most telling description of disenchantment is found in his "Science as a Vocation," where he writes that "there are no mysterious incalculable forces that come into play, but rather that one can, in principle, master all things by calculation. This means that the world is disenchanted. One need no longer have recourse to magical means in order to master or implore the spirits, as did the savage, for whom such mysterious powers existed. Technical means and calculations perform the service" (Weber 1991 [1919], p. 139). In other words, disenchantment is the turn to a belief that what we call rational and logical thinking can deliver us all the possible information about reality, so that there is nothing in reality that requires non-rational and non-logical insights. Disenchantment is closely connected with processes of secularization and rationalization. As Richard Jenkins (2000) sees it, secularization is the more negative process of neglecting magic and religion as well as the values and beings connected with them, while rationalization is a more positive process of implementing formal procedures into an expanding field of human activity. Weber himself—as well as most of his followers—saw the process of disenchantment—together with processes of secularization and rationalization—as an ambiguous, double-edged sword. On the one hand, it is a liberation from the cage of beliefs in supernatural forces and a chance for humans to take matters into their own hands. But on the other hand, another prison has appeared because—to invoke another of Weber's celebrated metaphors—the iron cage of bureaucracy and other formal procedures grows, while the world is emptied of meaning and humans become alienated. The disenchantment of the world is thus an intrinsic element of modernization, and to some degree it can be said that the world is modern because it is disenchanted.

Many accounts of modernity and disenchantment offer a one-sided view. One type assumes that the disenchantment of the world is a progressive process that increasingly covers more and more of the "Western" world. Morris Berman (1981) offers one of the most thoughtful and exhaustive examples of this approach. For Berman if we look at modernity, and even earlier periods of Western culture, we can see a history of consistent disenchantment of the world without any significant alternative processes. He believes that it has resulted in alienation from other people, detachment from nature, erosion of meaningfulness and neurotic anxiety. After examining the disadvantages of disenchantment of the world, he postulates the need for a re-enchantment. He expresses disenchantment with disenchantment, in the sense of being disappointed with its consequences and he is one-sided since he perceives modernity as containing only continous disenchantment

Another type of one-sided view of modernity and disenchantment suggests that there has been no modernity or disenchantment at all. Probably the most radical and most famous expression of this approach is that of Bruno Latour (1993) who claims that "we have never been modern" and therefore have also never been disenchanted. Probably the simplest way Latour's main argument can be summarized is to say that the disenchantment has taken place only at the level of theory and declarations, while actual practices have remained enchanted in the sense that social practices ignore modern dualism and consequently produce what Latour calls "hybrids"—beings that do not fit to modern classifications. Indeed, Latour claims that the modern work of purification (distinguishing and separating, particularly, but not only, what is human from what is non-human and what is cultural from what is natural) stimulates the proliferation of hybrids—although, at the same time, modernity pretends that hybrids are at best a rare exception and that, in general, everything fits into modern dualisms and categories. One of the most famous parts of his argument is the concepts of the agency of things, which refers to the idea that what moderns would like to consider as inanimate objects that are simply used or created by the human subjects, actually have also an impact on humans and in that sense could be considered as subjects as well— eventually, the difference between subjects and objects is not as clear and neat as moderns would like to see it since both humans subjects and non-human objects are hybrids that interact with each other. While I generally agree with the statement that we have been modern only in theory, but not in practice (although I am also aware that it might be considered an oversimplification or misinterpretation of Latour's meandering account), I disagree with his more general and radical claim that we have never been modern and disenchanted at all. The fact that disenchantment and modernity were not totalizing and took place only at some levels and not in others does not mean that they did not take place at all.

I would like to put aside radical accounts of disenchantment such as those developed by Berman (we have only been progressively disenchanted) and Latour (we have never been disenchanted at all)—not because they are false in general, but because they interpret the facts I generally agree with in a one-sided way and produce equally one-sided ideas. Instead, I will focus on more subtle and moderated accounts of modernity, disenchantment, and re-enchantment.

One type of moderated account makes the argument that disenchantment is indeed generally occurring, but that it is unable to eliminate all of the magic and enchantment. Jane Bennett's (2001) work is an example of this approach. Disenchantment, for her, has never been total, and she not only identifies examples of enchantments, particularly aesthetic ones, that persist, but also emphasizes their role in ethical considerations. Another kind of moderate approach is based on the premise that disenchantment transforms, rather than eliminates, magic or religion.

Bronislaw Szerszynski (2005) sees disenchantment as another level of the evolution of the sacred. The postmodern sacred, in his view, is private rather than public, individual rather than collective, pluralistic rather than monotheistic, and based on subjective experience rather than on inter-subjective beliefs about the objective order. Disenchantment is therefore not about the elimination of sacredness, magic, and so on, but about transforming them.

One of the most radical critics of the successfulness of disenchant-ment is Jason Josephson-Storm (2017), who—less famously, but more vlearly than Latour—claims that the effectiveness of disenchantment is so limited (because magic commonly prevailed in Western modernity even among intellectual elites) that our belief in disenchantment is a kind of a myth itself—in that sense he believes that he disenchants the dis-enchantment. What I find particularly plausible in his informative and comprehensive approach is that he suggests that we should speak rather about "disenchanting" than "disenchantment" to emphasize the ongo-ing, non-final and non-total character of this process. Actually I believe that the same refers also to the process that we call re-enchantment and which could be better labeled as re-enchanting.

It is also worth calling attention to the complex and sophisticated approach developed by Wouter Hanegraaff (2003). Hanegraaff claims that if we believe that science or art can adapt to new social and cul-tural conditions, then there is no reason to think that magic lacks this ability and socio-cultural transformations must result in its disappearance. He therefore claims that the disenchantment resulted in the survival of magic in the form of "disenchanted magic." Such magic seeks a justifica-tion not in religion (as, Hanegraaff argues, Renaissance magic did), but rather in science, particularly in psychology. It is closely connected with a shift mentioned by Szerszynski as well: from objective truth to subjec-tive experience. Another aspect of it, in Hanegraaff's view, is a shift from finding sacredness not in a transcendent god but in the immanent world of nature.

The notion that magic survived the disenchantment of the world leads Hanegraaff to a claim that we now live in two separate worlds: the disenchanted one in which we act and think in a rational and log-ical mode, and the world of (disenchanted, but still) magic in which modern rationality and logic are put aside. Following Lévy-Bruhl and Tambiah, Hanegraff distinguishes two "spontaneous tendencies" that both "belong to human nature": participation and instrumental causality. He argues that they functioned alongside each other in Western (as well as other) societies until the disenchantment of the world led to pressure to elevate the status of instrumental causality and depreciate the status of participation. Participation and magic, though suppressed, nonetheless survived by adapting to the new conditions and creating a new, separate world. Hanegraff contends that "not only does the feeling of participation explain the continuous attraction of magic in a disenchanted world—the experience of disenchantment actually causes an emotional need to reaffirm participation" (2003, pp. 377–378). In other words, magic survives or even flourishes and a kind of re-enchantment takes place not in spite of disenchantment and rationalization, but because of them. This seem to be a plausible explanation of the increasing tendency to re-enchantment, which expressions and examples are broadly discussed by Christopher Partridge (2004, 2005), among others.

There are other accounts that also suggest that disenchantment leads to re-enchantment. Zygmunt Bauman (1992) argues that while modernity attacked premodern illusions (magic, religion) by disenchanting the world, postmodernity attacks the modern illusion that everything can and should be disenchanted (be rationalized and rendered calculable), which has led to emptying the world of meaning and other negative consequences by re-enchanting it (Bauman 1992). Edward Tiryakian (1992), meanwhile, believes that disenchantment and differentiation, on the one hand, and re-enchantment and de-differentiation, on the other, are dialectical processes that are characteristic of modernity. Finally, Georg Ritzer (2009) has famously suggested that in contemporary consumer culture magic is often intentionally and rationally designed and that—to oversimplify—re-enchantment is actually *used* by disenchantment. He discusses examples of what Jean Baudrillard called "implosion," the blurring of boundaries, for example, between real and unreal—obviously, this process can be seen as similar—if not the same—as the de-differentiation described by Tiryarkian.

All these moderated accounts briefly mentioned above not only point out to the enchantments that survived disenchantment, but also to the increasing process of re-enchantment and moreover they claim that the re-enchantment is actually a result of disenchantment. Therefore these accounts show disenchanting and re-enchanting as complementary processes that define modernity.

It is important to acknowledge, however, that this general picture of disenchantment (and differentiation) and re-enchantment (and de-differentiation) coexisting socio-culturally is also visible in a more "local" field of interactions with technology. Actually, some users of new technologies explicitly connect their experiences with it and attitudes toward it in terms of magic and enchantment by calling themselves technopagans or technoanimists (Aupers 2002). I described the way magic can be seen in interactions with robots in the previous chapter. Others argue, however, that all new technologies tend to head toward re-enchantment (Kamińska 2007; Stivers 2001). On the other hand, Nicolas Carr's and Sherry Turkle's analyses of internet use and automation may be interpreted as descriptions of the role of technology in disenchanting the world, reducing it to a set of rational procedures and algorithms that can be performed by robots or robotized people. Indeed, even the "glass cage" metaphor used by Carr seems to hark back to Weber's account (Carr 2011, 2015; Turkle 2012, 2015). Finally, Mark Coeckelbergh shows affinities with Bauman, Tiryakian, and Ritzer in the way he sees the presence of both rational and romantic aspects in our interactions with and attitudes toward technology as two sides of the same modern coin (Coeckelbergh 2017).

I would like to suggest that re-enchantment is not only a matter of magic or other enchanted elements of reality developing a renewed presence as a kind of reaction to, or against, disenchantment, but also an expression of changing evaluations of its presence. Although seeing magic as a relic of a primitive and childish past or as symptom of mental issues still persists, as I argued in the previous chapter, there is a strong and growing counter-tendency, as well: magic is increasingly perceived as acceptable, normal, and natural, even as an enriching and necessary part of human existence. Therefore, if we follow Hanegraaff's understanding of disenchantment as pressure to suppress or eliminate magic, re-enchantment can be understood as an opposing tendency to express and celebrate magic.

All of the foregoing examples and diagnoses of re-enchantment (and disenchantment as well)—at both the global level of socio-cultural transformations and the local level of interactions with technology—relate to the level of engaged interacting with the world in general or technology in particular. But the level of thought and reflection about the world of technology, where the coexistence of disenchantment and re-enchantment is also at work, is equally valuable to consider. This is why I believe that Gunkel's and Coeckelbergh's ideas about how to think about our thinking about robots express a broader shift in our thinking about the presence and status of magic and magical thinking. Their ideas—as I argued in Chapter 3—share significant similarities with magical thinking and explicitly suggest the need to draw inspiration from premodern thought.

It is not surprising that the field in which this tendency is probably most visible is anthropology. Representatives of this discipline increasingly suggest that we ought to use ideas, perspectives, and methodologies for a long time ascribed to Others and used to describe them (i.e., "natives", "pre-moderns") in order to describe ourselves and our world, as well. Such tendencies are visible in the indigenous research remarked in Chapter 3, but there are also other, less explicit reconsiderations of concepts connected with magic and magical thinking. Animism, in particular, is a concept that has recently experienced a renaissance. Timo Kaerlain (2015) calls this tendency "neoanimism." Tim Ingold (2006) suggests that it may be a way to re-animate our thought. I believe this re-animation has two main aspects: an epistemological aspect (often coming together in a complementary way with animistic ontology) and an ethical aspect (although it should probably be rather called axiological, since ethics seems to be a rather anthropocentric term that refers mainly to human beings, while re-animation brings other entities in and even displaces people as subjects).

With the epistemological (and ontological) aspect, the main concern is focusing on relations rather than on objects. Nurit Bird-David (1999), in revisiting animism, speaks about relational epistemology, which has significant similarities with the parts of Gunkel's and Coeckelbergh's ideas that have been considered part of the relational turn. Ingold also highlights the importance of relational thinking and adds to it the primacy of movement. Neoanimism thus postulates an epistemology (and, at least to some degree, an ontology) that at its center has relations instead of objects, processes and events instead of things, and verbs instead of nouns. In fact, this dynamic vision of reality is part of premodern, magical worldview, as I argued in the previous chapter.

Alf Hornborg (2006) similarly believes that relationism is the only reasonable alternative to objectivism and relativism. He rightly emphasizes that when it comes to contemporary reevaluations of our epistemologies, methodologies, and ontologies, there is also a tendency—he mentions Latour and Ingold as its representatives—to eliminate or at least problematic such modern dualisms as subject/object (putting an emphasis on relations is in fact a way of blurring this distinction) and culture/nature (Latour's emphasis on the presence of hybrids tends to show that there are, at best, very few things that are purely cultural or natural.)

We can point out again here that a similar tendency is found in more local thinking about robots and technology. It is no accident that each time I mention neo-animistic epistemology I also make a remark about ontology. Epistemologies based on premodern animism are not neutral vessels that produce pure theory detached from reality, nor are they involved more in questions about how words relate to things than in presenting an ontology and saying how things are. They are involved in the world and have a strong ontological vision of it. This shift from epistemology to ontology seems to be part of a broader trend in contemporary humanities and social sciences, including speculative realism in philosophy, for example.

Finally, neoanimism proposes a shift in the very general attitude toward the world when we try to understand it. Tim Ingold and Mark Coeckelbergh, for example, suggest more engagement and involvement and less detachment and distance, in addition to connecting knowing with being, and thinking with living. In other words, it seems that the point here is to understand the world not through distancing oneself from it, but rather by being involved in it.

As I have already mentioned, this engaged epistemology and relational ontology have a strong axiological aspect. Neoanimism—which can be seen as a part of a non-anthropocentric, posthumanist axiology—seeks to reevaluate the attitude of human beings toward the rest of the world, to move from perceiving everything non-human as a resource and a tool to a more holistic and non-anthropocentric view in which humans and other beings are equal elements of intrinsically interconnected environment. Graham Harvey succinctly articulated this egalitarian approach in a book, titled significantly, "Animism: Respecting the Living" (2005).

Similarly, at the level of reflection about the world, re-enchantment can be seen as a result of disenchantment. One way such disenchantment can be seen in the work of certain economists, who—according to Michael Sandel (2012), for example—tend to reduce all human-human interactions to rational transactions by stating that economics is not only about what happens in the market, but about each and every interaction that takes place between humans. Another example is the reductionist approach taken in some of the natural sciences, which, in

Andrzej Zybertowicz's (2015) opinion, claim that everything that people do can be reduced to chemical and/or biological processes, particularly those that happen in the brain. These two examples are, as I have been suggesting, part of a broad tendency of disenchantment at the level of reflecting about the world that results in the world being perceived mainly as a set of individual subjects that make rational transactions with other subjects and use everything as a commodity, while in fact all of this can be reduced to natural processes. The re-enchantment of the world at the level of reflecting about it can therefore be seen as a result of disenchantment taking place at the same level, and as resistance to a cold, disenchanted vision of the world, where everything is rationalized, quantifiable, and described by abstract theories detached from lived experience. Examples of re-enchantment suggest that we should look for better—in both the epistemic and the ethical sense—ways of interacting with and reflecting on the reality.

To summarize this section, I would like to reiterate that its main purpose is to sideline one-sided visions of modernity that see modernity as involving only progressive disenchantment, so we have only been disenchanted, or that we have never been neither disenchanted nor modern at all. Instead, I have highlighted accounts that understand modernity as a matter of a coexistence of—opposite but complementary—processes of disenchanting and re-enchanting of the world.

I have been attempting to shine a light on various instances where we can observe the modern coexistence of the disenchantment and re-enchantment of the contemporary world. Re-enchantment can be discussed both globally as a process occurring in the whole of Western culture or more locally, in a specific field of socio-cultural reality. The presence of magical thinking in interactions with robots, as discussed in the previous chapter, is an example of such a local aspect of re-enchantment.

Re-enchantment can also be discussed at the level of interactions with the world as well as at the level of reflection about the world. The former can be found in analyses that point to the increased global or local presence of magic and enchantment in the spontaneous actions and interactions of individuals involved in the world. The latter is present in manifestos that call for reshaping and reevaluating the perspective from which we perceive reality, our way of thinking about the world. Put differently, the first level is about the re-enchantment already happening in our world; the second level is about proposals to change how we think to bring it closer to the ways of thinking present in the enchanted, non-modern world.

Re-enchantment might be more or less aware and intentional, or unaware and unintentional; it is probably most often intentional. Think of all the new spiritual approaches that explicitly refer to magic. Neoshamanism is an example. But think also of the way consumer goods are designed to make them magical and enchanting. The latter type is less common and less obvious since it refers to situations such as interactions with robots when magical thinking is done without any conscious intention to do it and without being aware of it.

In sum, I disagree with both arguments: those that claim disenchantment has never occurred at all, and those that emphasize the totalizing, unlimited, and undisputed character of disenchantment. My own view is that disenchantment is limited in the sense that not everything in the world has become disenchanted, and that some things that have become—at least to some degree—disenchanted, have become re-enchanted as a result. I would argue that not only has some magic avoided being disenchanted, but also that disenchantment has led to re-enchantment, to a stronger presence and more positive evaluations of magic. In the next section, I show how exactly is this happening by identifying functions of re-enchantment and magic that compensate for the negative consequences of modernity in general and disenchantment in particular.

4.2   Re-enchantment and Its Functions

In this section, I examine the claim that re-enchantment is a result of disenchantment. I look, in particular, at the functions of re-enchantment that make up for some things that are lacking as a consequence of rampant disenchantment. I discuss it at the global level of general socio-cultural transformations and at the local level of attitudes toward technology on the one hand, and at the level of interaction about the world and reflection about the world, on the other. I also make use of the socio-regulative theory of culture to illustrate the above-mentioned processes in more abstract terms. Throughout this whole section, I emphasize that re-enchantment is not a simple return to the past, but rather needs to be understood as a new and specific phenomenon taking place in context that is, in many ways, different from how the world was before disenchantment: contemporary magic is significantly different from the magic of "magical societies".

The first factor that may determine the re-enchantment and the increased presence of magic and magical thinking might be lack of understanding. As for the level of interaction with the world, it refers to the global situation in Western culture in general and to interactions with robots in particular. It is very often stated that the world changes so quickly that it is difficult to keep up and that such situation is partially due to technological progress (Agger 2016; Gleick 2000; Wajcman 2014). The world is changing at such a fast pace that it is hard to understand it. Most people know how to send a text or make a call, but they do not know how to explain how mobile phones actually work, and the same goes for a significant number of objects and mechanisms that are present in the contemporary world. Robots are another example of objects that most of us could operate, but the engineering that goes into them is understood only by a relatively small minority. Our world is full of new objects and events—including robots, that we do not rationally understand (of course it is possible to understand them, but that takes a lot of time and effort that we cannot or do not want to make). In order to make some sense of it all, we retreat to the most basic and fundamental way of understanding—that is, to magical thinking. As I pointed out in the previous chapter, anthropologists and psychologists have provided strong evidence that magical thinking and magic are the most basic and primal way of understanding reality, particularly when the object of understanding is for some reason difficult to be grasp by "rational" and "logical" means. We are like children who appear in a world in which everything is constantly new and surprising. Magical thinking is the easiest way to achieve a basic understanding. Lack of understanding is closely connected with a lack of security and anxiety; when we do not understand the world, we feel insecure and anxious.

Lack of understanding is also present at the level of intellectual reflection about the world. Andrzej Zybertowicz (2015) claims that due to the unprecedented pace, extent, and depth of socio-cultural transformation—particularly in the sphere of technology—we not only increasingly fail to understand the world, we also increasingly fail to care to understand the world. He believes that—somewhat paradoxically—we leave the task of understanding to technology and often do not make an effort to interpret the raw data provided by it. Instead, we simply assume that the quantitative results of technological calculations are the only kind of knowledge we need. Zybertowicz calls this the suicide of the Enlightenment, since he considers the whole process not a result of re-enchantment or an increasing role for irrational factors, but rather as a result of rampant disenchantment and an expansion of the instrumental reasoning that reduces everything to numbers and calculations (Zybertowicz 2015). I believe that the thinkers I have discussed in this and the previous chapters such as Cockelbergh, Gunkel, Ingold and Latour (my recognition of magic and magical thinking in Latour's thinking and my understanding of it as an expression and recommendation of re-enchantment is inspired by Michał Rydlewski [2014] and Andrzej W. Nowak [2016]) who can be read as proponents or even providers of a re-enchantment of the world at the level of reflection about the world, are this partially because they think that our modern epistemology and modern knowledge do not give us a proper understanding of contemporary reality. Modern dualism does not fit the increasingly hybrid reality (if it ever fit reality at all), and rational and logical thinking does not account for all of the syncretic elements of reality. The lack of understanding at the level of reflection about the world (comparable to the level of interacting in reality) leads to a lack of security and control and, in consequence, to anxiety.

The lack of security and control as well as the feeling of anxiety are a second factor that is shaping the re-enchantment of the world and the increasing presence of magic and magical thinking. As for the level of interactions with the world, it is well known that a significant number of sociological investigations of contemporary Western culture find that we live in a "society of risk" (Beck 1992), that our life is "liquid" (Bauman 2000), that we experience a lack of ontological security (Giddens 1992) and increasing precariousness (Standing 2011). In the chapter "Robots Enchanting Humans" I have showed how important this issue is in the context of intimate relationships. The contemporary world is not only difficult when it comes to understanding it, but also when it comes to feeling at home in it—a home that is to some degree predictable and controllable. Without predictability and security, we feel anxious. Technology and robots work in a comparable way and seem to even increase the level of insecurity and anxiety, due, for example, to the specific scenarios and narratives we develop about the robots in Western culture, and which generally stimulate the fear of robots replacing humans. These scenarios are as follows: the first one refers to machines taking control over people, the second refers to robots taking jobs and leaving more and more people unemployed without a main source of income, and the third one refers to social robots that will replace humans in intimate relationships or at least impoverish them (of course there is also the second side of the coin: some would claim that the second and the third scenario should be a cause for enthusiasm rather than skepticism, insecurity, and anxiety). Insecurity and anxiety in interactions with robots are thus connected with the fear that robots will replace human beings in professional tasks, intimate relationships, and will ultimately take control over people. We should note, as well, that interactions with robots may involve an "uncanny valley" phenomenon, which has been famously described by Masahiro Mori as resulting in a feeling of eeriness. The uncanny valley might thus be a source of additional anxiety (Mori 1970). As I pointed out in Chapter 3, the majority of studies of magical thinking—both theoretical and empirical, anthropological, and psychological—emphasize that its main function is providing a sense of security and control as well as relief from anxiety.

Insecurity and anxiety seem also present at the level of reflecting about the world. Alf Hornborg (2006) claims that "it is the predicament of modern people to remain chronically uncertain about the validity of their own representations" and labels this "modern condition" as "reflexive uncertainty" (p. 27). Although this statement refers both to the level of interacting with the world and reflecting about the world, I would like to focus on the latter one. Modern philosophy, at least since Descartes, and due to other thinkers such as Kant, is chronically skeptical and uncertain about its own statements—it seem that it emphasizes more negative qualities than positive ones, and focuses on showing more what we cannot know than what we can know.

A lot of twentieth-century humanities and social sciences—particularly anthropology—are influenced by this deep skepticism, which has been further stimulated by postmodernism. This "reflexive uncertainty" involves a decreasing role of ontology and an increasing role of epistemology, since we first need to understand the way we perceive the world and find a good way to do that before we will say anything about the world itself. I believe that increasing numbers of philosophers and other humanities scholars feel anxious that this epistemological carefulness in fact prevents us from talking about the world itself. They feel anxious because the knowledge that their disciplines produce seem to be increasingly detached from reality, from practice, from life. One of the expressions of this tendency is speculative realism with probably most radical representative, Quentin Mellasoux, who wishes to eliminate "correlationism" and who brings back the view that we can represent reality as it is, that we can access the things themselves, not only the phenomena but also nou-mena. Meillasoux thus attempts to overthrow epistemological uncertainty and carefulness (Meillassoux 2010). This ontological turn is often seen as inspired by Bruno Latour, who seems to share the anxiety and impatience about the abstract and detached character of modern dualisms that do not fit and do not account for what actually happens in practice. He also sig-nificantly criticized postmodern philosophy for a comparable—but differ-ent in many ways—detachment from practice and reality.

As for the philosophy of technology, something similar seems to take place, although it takes a little bit different shape. David Gunkel and Mark Coeckelbergh emphasize the fact that our ideas about the robots do not fit our actual experiences of them. They both emphasize that our knowledge is detached from what is actually happening at the level of interacting with the world. Mark Coeckelbergh suggests very succinctly that we need to develop knowledge that provides understanding is less distance and more engagement. David Gunkel puts a lot of effort into showing how narrow modern dualisms are and how much they limit what and how we can think—primarily on the ethical level, but also more generally. In this sense, contemporary thinkers seem to feel anx-ious about the "reflexive uncertainty" of modern thought; they see it as resulting in a detachment of the thought from reality, from practice and from actual experience. The direction in which they are heading to change this state of affairs is a kind of re-enchantment of the world, which in fact seems to fit their aims since magical thinking is probably the kind of thinking that is as embedded in the world, practice, and experience as strongly as possible.

Finally, a third possible cause of the contemporary presence of magical thinking is lack of meaning. Max Weber (1991 [1919]) underscored this consequence of disenchantment in his seminal diagnoses: "Precisely the ultimate and most sublime values have retreated from public life either into the transcendental realm of mystic life or into the brotherliness of direct and personal human relations" (p. 155). As I tried to show in the chapter "Robots Enchanting Humans", the situation has significantly changed, since—due to the disenchantment of humans and disenchant-ment with humans—"values have retreated," at least partially, from the intimate relationships as well. Therefore, if the last bastion of meaning is "realm of mystic life" than re-enchantment seems to increase this realm and enable the return of meaning. Nevertheless, it needs to be empha-sized that this meaning is no longer embedded in values, but rather in emotions and experiences. The main difference is that, while values are intersubjective, emotions and experiences are subjective. While values demand articulation and common understanding to provide meaning, emotions and experiences may provide individuals with meaning regardless of whether anyone else understands or is interested in them. Both Szerszynski and Hanegraaf highlight this subjectification, individualization, and internalization of the experience of magic. Moreover, Ritzer's analysis of consumerism also seems to confirm that what is enchanting in the way that commodities are presented as a spectacle and in the participation in that spectacle is not any kind of common, intersubjective value that demands articulation and common understanding, but rather a personal, subjective experience, a thrill, a sensation that requires neither words nor understanding.

Finally, as I have elaborated in the chapter "Robots Enchanting Humans", some expectations about intimate social robots (lovebots, sexbots, and other artificial companions) are also focused on the opportunity of gaining strong and positive emotions and experiences from interactions with them without limitations or any of the unpleasant elements that are commonly experienced in similar interactions with human beings. One of the advantages of intimacy robots—according to their enthusiasts—is that there is no need to make any agreement with robots, to develop consensus about beliefs or values. An individual may focus on one's own subjective emotions and experiences without being limited by the other side of the relationship. It is also worth noticing an ambivalence in our attitudes toward robots. On the one hand, robots are becoming a promise of a new, fulfilling life that will finally make sense, providing their users with security, positive experiences and freedom. On the other hand, robots still remain a source of insecurity and anxiety since they threaten to take away our jobs, and to outgrow us in various ways, eventually becoming something better than human beings. It therefore seems that re-enchantment is a reaction to a lack of meaning, although the meaning comes back in a different form, one of subjective emotions and experiences rather than inter-subjective values.

As for the level of reflection about the world, it seems that lack of meaning is expressed toward the status of knowledge produced from the modern perspective. Such knowledge—as I have already noted—does not provide an understanding of our reality, is detached from practice and experience and, therefore, seems to be meaningless since it does not help anyone neither to adapt to the reality nor to change it into a better one. Moreover, when modern knowledge turns out to have meaning and to emphasize values, it often turns out that these are in fact meanings and values we no longer consider precious—in other words, when it provides a logocentric and anthropocentric expansion of instrumental and economical reason that promotes thinking of everything and everyone in terms of objects, tools, and resources and eventually leading to degradation of environment and erosion of interactions with other living beings. Thinkers like Latour, Ingold, and Coeckelbergh try to imbed the knowledge in practice, reconnect knowing with living and provide ways of thinking that would enable more holistic and egalitarian ontology, epistemology and axiology. In this sense, Tim Ingold—as I pointed out above—claims that we need to reanimate our thought.

To summarize what has been already stated in this section, it can be said that the re-enchantment takes place as a compensation for three lacks resulting from rampant disenchantment: a lack of understanding, a lack of security and control, and a lack of meaning. These compensations can be observed both at the level of interaction with the world, and reflection about the world, and our thinking about robots and our thinking about our thinking about robots (described in the previous chapter but recalled in this section) are significant expressions of them, respectively. It therefore seems that nowadays re-enchantment and magic perform similar functions that magic has been performing before (traditional) religion, art, and science appeared. It provides security, meaning and understanding, while (traditional) religion, art, and science fail to do this due to their rampant disenchantment. Below, I would like to put more light on these issues (as well as on some diagnoses made in the previous section of this chapter) by referring to the socio-regulative theory of culture developed by Jerzy Kmita.

Jerzy Kmita distinguishes two main spheres of culture: symbolic and instrumental. The difference is that actions regulated by beliefs from the instrumental sphere of culture do not need to be commonly respected to be effective, while actions regulated by beliefs from the symbolic sphere do. In other words, the symbolic (sphere of) culture involves interpretation and the common respecting of particular beliefs as a necessary condition that enables others to understand us and our actions; instrumental (sphere of) culture does not involve interpretation or—to be more precise—the interpretation of our actions does not determine their efficiency and effectiveness, which is actually determined by objective conditions. Another way to say it is that actions regulated by symbolic culture require intersubjective consent, while actions regulated by instrumental culture do not.

Nevertheless, Kmita claims that the distinction between those two spheres of culture is not universal, and in fact it mainly fits the cultures of "modern" societies. In his opinion, the cultures of other societies, particularly "magical" societies, are structured differently. Kmita argues that magical cultures are—as I made clear in the chapter "Humans Enchanting Robots"—originally syncretic: they do not involve the distinctions that later types of societies do, which comes from the fact that there is no distinction between symbolic and instrumental culture. I will now briefly present Kmita's and Pałubicka's vision of how the structure of culture has evolved from the "magical societies" (with the originally syncretic magical culture typical of them) to industrial societies (with the disenchanted culture typical of them) and add another approach that grasps the process of re-enchantment described in this chapter.

In the magical cultures, there is no distinction between instrumental and symbolic spheres of culture (nor between materiality and spirituality, mentality and physicality, and so on)—since the distinctions that we take for granted from the point of view of our culture have not developed there (yet), or their presence is very limited. Following Pałubicka, we can call this state of culture "original syncretism." While syncretism is obviously a term that refers to the mixed character of spheres of culture (and—to simplify—almost everything else), its originality means that this mix is not a result of a mixing of something that was separate before, but rather is a consequence of the fact that the separations and distinctions have not been made yet. Nothing is simply instrumental or symbolic: everything is both—this state of affairs is illustrated below with Fig. 4.1. As I pointed out earlier, this approach derives from positions such as those of early Lévy-Bruhl, who claimed that for "magical societies", nothing is natural or supernatural, animate or inanimate, and in some sense everything is equally mystical and equally real. That is why when "magical societies" prepare to go hunting (which from our perspective is regulated by instrumental culture) rituals and prayers are equally important as sharpening the spears and arrows (due to modern beliefs, hunting has instrumental character, so only the latter preparation is justified since only it has anything to do with effectiveness of the activity). Or the name (purely symbolic from modern point of view) is treated as an extension of its owner, and saying it in some specific contexts may have concrete results for the situation of the owner, regardless of the fact of whether the owner is aware of the situation (which is quite absurd from modern perspective, since it perceives the act of saying someone's name as purely symbolic, and to have any effect on someone, the communication needs to be heard and interpreted by that person).

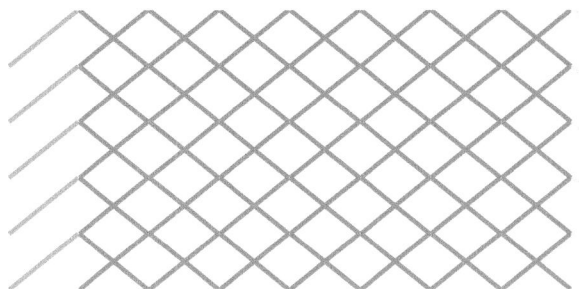

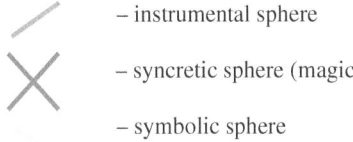

– instrumental sphere

– syncretic sphere (magic)

– symbolic sphere

Fig. 4.1 Original syncretism: enchanted world

According to Kmita and Pałubicka, the development of culture, described as the disenchantment of the world, can be understood as a gradual separation and autonomization of the symbolic and instrumental aspects of culture. As a result, the culture of "modern" societies of the first half of the twentieth century can be characterized by a rather clear distinction between two spheres of culture and an autonomous character that entails taking "modern dualisms" for granted. It can be pointed out, of course, that these distinctions are implemented only in theory, but practice often remains largely syncretic, hybridic, and thus magical. Even if this is the case, however, it is still a significant transformation in comparison with "magical" societies which—according to this approach—acknowledge these distinctions neither on the level of theory, nor at the level of practice. So, the structure of culture in the "modern" societies is presented by the picture below: we can see that symbolic and instrumental culture are to a large degree distinguished from each other, and that syncretic magical mix of them ("disenchanted magic" in Hanegraff's terms) is a relatively small part of the whole structure—this situation is illustrated with Fig. 4.2.

It is now time to use socio-regulative theory to illustrate and explain disenchantment and re-enchantment in modernity, to further develop the examination presented in this chapter. Kmita and Pałubicka suggest

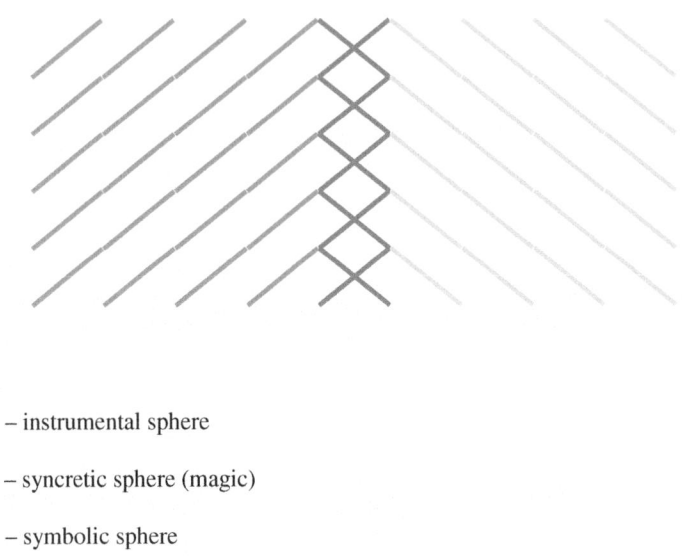

– instrumental sphere

– syncretic sphere (magic)

– symbolic sphere

Fig. 4.2 Autonomy of spheres: disenchantment of the world

that the further process of disenchantment might involve desymbolization. This means increasing the role of instrumental culture and the decreasing the role of symbolic culture. Put differently, the actions (or classes of actions) regulated by symbolic culture would become regulated by instrumental culture. In my previous book, I interpreted Habermas's concept of a colonization of the Lifeworld that reads this process as an example of desymbolization. In fact, Habermas claims that colonization entails a replacement of communicative actions that aim to achieve some intersubjective consensus with instrumental actions that aim to achieve instrumental goals, most often without the necessity of achieving consensus (Habermas 1985a, b; Musiał 2015). Even without any detailed elaboration, it is quite clear that Habermas's instrumental and communicative actions are to a large degree comparable to Kmita's instrumental and symbolic culture and that, in consequence, colonization of the Lifeworld can be considered an expression of desymbolization.

The rationalizing and commercializing tendencies in intimate relationships described by Illouz and Hoschschild and mentioned in the chapter "Robots Enchanting Humans" are also examples of this process.

Technology is another example—when Nicolas Carr points out that the information we find on the internet is largely left without interpretation and that its processing is shallow and robot-like, it seems that he is also analyzing desymbolization. Many more examples of tendencies of desymbolization in contemporary culture are discussed by Jarosław Boruszewski (2015).

Generally, I would see as examples of desymbolization most of the analyses that posit the increasing domination of what is most popularly called instrumental reason and the increasing role of quantitative data, as well as a growing tendency toward the rationalization and professionalization of interactions with other humans beings by implementing formal procedures for them. It is also crucial to notice that desymbolization is strongly connected with individualization, that is, with a tendency to emphasize individual autonomy and individual worldviews that define each of us as different than the rest. The stadium of rampant disenchantment that involves desymbolization is presented on Fig. 4.3: the symbolic sphere is decreasing, while the instrumental sphere is growing.

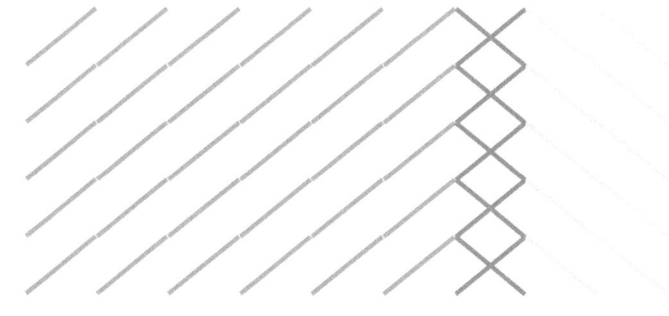

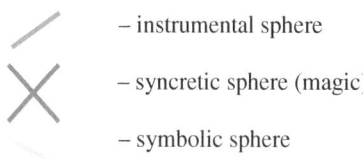

– instrumental sphere

– syncretic sphere (magic)

– symbolic sphere

Fig. 4.3 Desymbolization: rampant disenchantment of the world

The rampant disenchantment—somewhat paradoxically—leads to establishing a complementary counter-process—the re-enchantment of the world. The world disenchanted by desymbolization, lack of understanding, security and meaning—becomes a cold world of procedures, objects and tools. This is where re-enchantment comes in—it is crucial to see that this does not bring back the instrumental and symbolic spheres of culture back into balance, but rather mixes them into renewed syncretism. This stage is presented on Fig. 4.4.

This renewed syncretism and "disenchanted magic" are—as it has already been pointed out in this chapter—in many ways different in comparison with original syncretism and original magic. Above all, the original magic in "magical societies" is intersubjective and based on collective beliefs considered to be objective truths, while disenchanted magic in contemporary Western societies is largely individual and based on subjective experiences and emotions. This is also a reason why re-enchantment is occurring, rather than re-symbolization. Symbols demands intersubjectivity, interpretation, and consensus to be an effective carrier of meaning, and it seems that inter-subjectivity, interpretation, and consensus are the very elements of life that contemporary Western societies try to leave behind. With desymbolization, Western societies lose meaning (values)

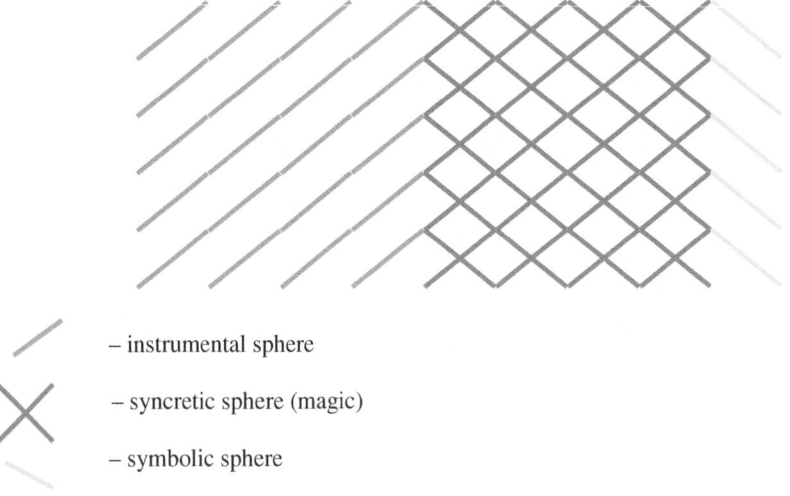

– instrumental sphere

– syncretic sphere (magic)

– symbolic sphere

Fig. 4.4 Renewed syncretism: re-enchantment of the world

and understanding, but thanks to re-enchantment, renewed syncretism, and disenchanted magic, they are able to bring the meaning back, but without the problematic and unpleasant burden of intersubjectivity, consensus and interpretation. They achieve them thanks to individualized magic, subjective experiences, magical understanding, which do not need to be articulated, interpreted, or acknowledged by anyone to remain effective.

4.3   Disenchanting Modernity

To speak about the disenchantment *of* modernity is to refer to our disillusions about the promises that are connected with modernity as well as to disappointment with its actual results. The disenchantment of modernity therefore means that that we simply no longer believe in modern reason, rationality, science and progress. We are skeptical toward the promise that they will enable us to understand the world or to make it a better place. But that is not all. There is also disenchantment *with* modernity, which means that we are not only disillusioned about promises of modernity. We are also strongly disappointed with the actual consequences of it: alienation from other people and from nature, treating everyone and everything as a commodity destined to be consumed, anxiety and lack of meaning—those, and many other disadvantages are seen as unpredicted and unintended, but also unavoidable and intrinsic results of modernity and disenchantment of the world. Re-enchantment is thus a result of disenchantment. Some of us want to romantically reclaim the magic that we have lost or to go beyond modernity, because we see that the promises of modernity and disenchantment have neither been fulfilled, nor can the actual consequences of them be considered positive. Of course, some of us still believe that modernity and disenchantment remain a valid, valuable project, and that despite many problems we should continue it and not let magic, re-enchantment and irrationality take control over us. Deciding who is right in this discussion is far beyond the scope of this chapter. Its aim was only to point out to the contemporary coexistence of disenchantment and re-enchantment and to show that the latter is a result of the former—particularly that disenchantment of and with modernity enables re-enchantment.

It is worth to notice that such disenchanting of modernity, which shows that we have never been as modern and as disenchanted as we thought we were and that we simply need magic and magical thinking, blurs the distinctions between "premodern" and "modern," between "magical societies" and children on the one hand, and "modern" "Western" societies on the other, just like the previous chapter has blurred the distinction between magic and rationality. I believe that the more we think about us and others, the more we might come up with conclusion that we have much more in common with each other than we thought.

REFERENCES

Agger, B. (2016). *Speeding Up Fast Capitalism: Cultures, Jobs, Families, Schools, Bodies.* Abington and New York: Routledge.

Aupers, S. (2002). The Revenge of the Machines: On Modernity, Digital Technology and Animism. *Asian Journal of Social Science, 30*(2), 199–220.

Bauman, Z. (1992). *Intimations of Postmodernity.* London: Routledge.

Bauman, Z. (2000). *Liquid Modernity.* Cambridge, UK and Malden, MA: Polity Press.

Beck, U. (1992). *Risk Society: Towards a New Modernity* (M. Ritter, Trans.). London: Sage.

Bennett, J. (2001). *The Enchantment of Modern Life: Attachments, Crossings, and Ethics.* Princeton: Princeton University Press.

Berman, M. (1981). *The Reenchantment of the World.* Ithaca, NY: Cornell University Press.

Bird-David, N. (1999). "Animism" Revisited: Personhood, Environment, and Relational Epistemology. *Current Anthropology, 40*(1), 67–91. https://doi.org/10.1086/200061.

Boruszewski, J. (2015). Symbolophobia and Pragmatomania. *Sensus Historiae, 19*(2), 39–53.

Carr, N. (2011). *The Shallows: What the Internet Is Doing to Our Brains.* New York: W. W. Norton.

Carr, N. (2015). *The Glass Cage: How Our Computers Are Changing Us.* New York: W. W. Norton.

Coeckelbergh, M. (2017). *New Romantic Cyborgs: Romanticism, Information Technology, and the End of the Machine.* Cambridge and London: MIT Press.

Giddens, A. (1992). *The Transformation of Intimacy: Sexuality, Love and Eroticism in Modern Societies.* Stanford: Stanford University Press.

Gleick, J. (2000). *Faster: The Acceleration of Just About Everything.* New York: Vintage.

Habermas, J. (1985a [1981]). *The Theory of Communicative Action, Volume 1: Reason and the Rationalization of Society* (T. McCarthy, Trans.). Boston: Beacon Press.

Habermas, J. (1985b [1981]). *The Theory of Communicative Action, Volume 2: Lifeworld and System: A Critique of Functionalist Reason*. Boston: Beacon Press.

Hanegraaff, W. J. (2003). How Magic Survived the Disenchantment of the World. *Religion, 33*(4), 357–380. https://doi.org/10.1016/s0048-721x(03)00053-8.

Harvey, G. (2005). *Animism: Respecting the Living World*. Kent Town: Wakefield Press.

Hornborg, A. (2006). Animism, Fetishism, and Objectivism as Strategies for Knowing (or Not Knowing) the World. *Ethnos, 71*(1), 21–32. https://doi.org/10.1080/00141840600603129.

Ingold, T. (2006). Rethinking the Animate, Re-animating Thought. *Ethnos, 71*(1), 9–20. https://doi.org/10.1080/00141840600603111.

Jenkins, R. (2000). Disenchantment, Enchantment and Re-enchantment: Max Weber at the Millennium. *Max Weber Studies, 1*(1), 11–32.

Josephson-Storm, J. A. (2017). *The Myth of Disenchantment: Magic, Modernity, and the Birth of the Social Sciences*. Chicago and London: University of Chicago Press.

Kaerlein, T. (2015). The Social Robot as Fetish? Conceptual Affordances and Risks of Neo-Animistic Theory. *International Journal of Social Robotics, 7*(3), 361–370. https://doi.org/10.1007/s12369-014-0275-6.

Kamińska, M. (2007). *Rzeczywistość wirtualna jako "ponowne zaczarowanie świata": pytanie o status poznawczy koncepcji*. Poznań: Bogucki Wydawnictwo Naukowe.

Latour, B. (1993). *We Have Never Been Modern* (C. Porter, Trans.). Cambridge, MA: Harvard University Press.

Meillassoux, Q. (2010). *After Finitude: An Essay on the Necessity of Contingency* (R. Brassier, Trans.). London: Bloomsbury Academic.

Mori, M. (1970). The Uncanny Valley. *Energy, 7*, 33–35.

Musiał, M. (2015). *Intymność i jej współczesne przemiany. Studium z filozofii kultury*. Kraków: Universitas.

Nowak, A. W. (2016). *Wyobraźnia ontologiczna. Filozoficzna (re)konstrukcja fronetycznych nauk społecznych*. Poznań: Wydawnictwo Naukowe UAM.

Partridge, C. (2004). *The Re-enchantment of the West: Alternative Spiritualities, Sacralization and Popular Culture and Occulture* (Vol. 1). London: T & T Clark.

Partridge, C. (2005). *The Re-enchantment of the West: Alternative Spiritualities, Sacralization and Popular Culture and Occulture* (Vol. 2). London: T & T Clark.

Ritzer, G. (2009). *Enchanting a Disenchanted World: Continuity and Change in the Cathedrals of Consumption*. Thousand Oaks, CA: Pine Forge Press.

Rydlewski, M. (2014). Kulturowy wymiar postkonstruktywizmu w perspektywie "Gramatyki kultury europejskiej" Anny Pałubickiej. *Filo-Sofija, 25*(2), 402–432.

Sandel, M. J. (2012). *What Money Can't Buy: The Moral Limits of Markets.* New York: Farrar, Straus and Giroux.

Sherry, P. (2009). Disenchantment, Re-enchantment, and Enchantment. *Modern Theology, 25*(3), 369–386. https://doi.org/10.1111/j.1468-0025. 2009.01533.x.

Standing, G. (2011). *The Precariat: The New Dangerous Class.* London, UK and New York: Bloomsbury Academic.

Stivers, R. (2001). *Technology as Magic: The Triumph of the Irrational.* New York and London: Bloomsbury Academic.

Szerszynski, B. (2005). *Nature, Technology and the Sacred.* Oxford and Malden, MA: Wiley-Blackwell.

Tiryakian, E. (1992). Dialectics of Modernity: Reenchantment and Dedifferentiation as Counterprocesses. In N. Smelser & H. Haferkamp (Eds.), *Social Change and Modernity* (pp. 78–94). Berkeley, CA: University of California Press.

Turkle, S. (2012). *Alone Together: Why We Expect More from Technology and Less from Each Other.* New York: Basic Books.

Turkle, S. (2015). *Reclaiming Conversation: The Power of Talk in a Digital Age.* New York: Penguin Press.

Wajcman, J. (2014). *Pressed for Time: The Acceleration of Life in Digital Capitalism.* Chicago: University of Chicago Press.

Weber, M. (1991 [1919]). Science as a Vocation. In H. H. Gerth & C. W. Mills (Eds.), *From Max Weber: Essays in Sociology* (pp. 129–156). Abington and New York: Routledge.

Zybertowicz, A. (2015). *Samóbójstwo Oświecenia? Jak neuronauka i nowe technologie pustoszą ludzki świat.* Kraków: Kasper.

CHAPTER 5

In Lieu of a Conclusion: Where Will We Go from Here?

In this book, I have attempted to understand the diverse ways in which we think about humans, robots, and the relationships between them, about magical and modern rational thinking, and, finally, about modernity. In doing so, I have focused on emphasizing and understanding the diversity of approaches that have been taken toward these issues, rather than on choosing the "right" one. I have explored what, how, and why we think in a certain way, again rather than deciding what kind of thinking might be correct. Nevertheless, it is difficult to escape the urge to ask who is right, since this question refers, in fact, to the future of the world in which we live. These considerations raise manifold questions. Should we welcome intimacy robots as an opportunity for more liberating, more secure, and more fulfilling relationships? Or should we ban such machines on account of the attendant risks of increased objectification, isolation, and deception? Will robots enrich our relationships or impoverish them? Should we try to re-enchant human beings as exceptional entities and to disenchant robots as tools controlled by people? Or should we instead seize the opportunity to eliminate human exceptionalism and consider ourselves as equal with the rest of the world, particularly by caring for our environment and granting robots rights? Whose welfare should we take into account? Should it be the good of individual humans, the common good of humankind, the common good of humans and robots, or the good of the environment as a whole? How should we conceptualize and measure the good that we take into

© The Author(s) 2019   141
M. Musiał, *Enchanting Robots*,
Social and Cultural Studies of Robots and AI,
https://doi.org/10.1007/978-3-030-12579-0_5

account? Should we continue to adhere to modern rational thinking and its dualisms, or should we try to think in other ways? Should we retain the modern order, try to correct it, or re-embrace the romantic notion of re-enchanting the world? Or should we try to identify some other path? Should we go back to recover what we have lost, go forward to find out what we can achieve, or fundamentally change our direction?

Once more, I have no easy or definitive answers to these questions; indeed, I assume that no such answers exist. My responses are difficult and contingent: in order to answer such questions, I suggest, we need to take into account all of the answers that are available in our contemporary culture and—before trying to decide who or what is right—seek to understand these answers by considering their origins, the tendencies that they express, and the cultural processes to which they belong. It is in these respects that this book is intended as a contribution to the effort to define the roles of humans and robots in the new society that is emerging.

In making the effort to understand our thinking, and our thinking about our thinking, though, I have come to wonder increasingly whether we overestimate its autonomy, transparency, and impact. It seems to me that we tend to forget that our thinking is often determined by factors of which we are not consciously aware. It is also my contention that we continue to assume that what we do is determined mainly by what we think. Neither of these statements is entirely true: the world is not entirely under the control of our beliefs and actions, nor are our beliefs or our actions purely the result of conscious reasoning. Instead, quite often, the world around us makes us believe in things and act in ways that are not the product of our conscious decisions or that are even contrary to those decisions.

One aspect of the world that has a powerful impact on us in these respects is technology. Though many of us have refrained from technological instrumentalism and embraced technological determinism, we still for the most part assume that we are in control of technology, able to decide what to do with it and what it will do with us. Thus, while Marshall McLuhan and other theorists have discussed the fundamental ways in which new tools and technologies may change our thinking, we nevertheless overestimate our impact on robots and technology and underestimate their impact on us.

Another aspect of the world about which we are significantly silent when discussing our thinking about robots, humans, and modernity is capitalism. We often seem to treat capitalism as something like the air that surrounds us, assuming that it has always been part of the fabric of our existence and always will be. Yet this particular approach to economics is embedded in a historical system that took shape at a particular moment in time and will presumably one day disappear or undergo significant transformation. In fact, capitalism may be fundamentally reshaped as a consequence of technological progress, for instance resulting in technological unemployment that necessitates the implementation of a universal basic income. We should, I suggest, seek to understand how capitalism shapes our thinking about technology in general and then try to think differently about technology and to form a prognosis regarding our future thinking about and interactions with technology. We must keep in mind that our thinking is embedded not only in modern rationality but also in capitalism.

There is a need, then, to recognize both the limits and the determinants of our thinking, a need to recognize that what happens in our world is not only a matter of what we think and that what we think is in part a matter of what happens in the world in which we live. That world and its enchanting, disenchanting, and re-enchanting parts—including humans, robots and modernity—are neither beyond nor entirely determined by our thinking and our actions.

It is also necessary, I am convinced, to give greater attention to the potential unintended consequences of the ideas, actions, and tendencies that we can already observe. That is, we need to stretch our collective imagination to its limits in order to think about the world that we might develop despite, rather than because of, our efforts to shape it. The purpose of such investigations would be both to prepare ourselves for a possible future and also to understand better our beliefs and desires and the cultural tendencies and processes at work in the world today.

I now offer examples of what I mean in the form of more or less critical discussions of the ideas of two philosophers who have attempted to envision—to echo the title of this chapter—where we can go from here. I begin with John Danaher's account of automation, achievement, and the meaning of life and then engage with Steve Petersen's arguments regarding artificial persons as servants.

5.1   Intimacy and the Meaning of Life in a World of Widespread Automation

The first example of future scenario I find worth considering due to the fact that it refers to unintended consequences of our contemporary ideas, actions and tendencies we observe refers to a widespread automation and the meaning of life. Widespread automation has been associated with technological unemployment, poverty, and increasing economic inequality. However, some philosophers have argued that a universal basic income or some other solution can solve these problems, after which the main problem for people would be finding meaning in life. Thus David Gunkel (2017), for example, argues that education must be rethought and reshaped to prepare people for a life of gainful unemployment. John Danaher (2017a, 2017c) similarly worries that, as robots replace human beings in more and more activities, the latter, no longer able to achieve anything without the aid of the former, will be transformed into passive recipients of benefits resulting from robots' achievements whose lives are devoid of meaning. He suggests that we might find the sense of achievement in playing games or using integrative technology e.g. making technology a part of us rather than an autonomous being. From my own perspective (Musiał 2018), one of the most interesting issues is the possible shift in attitudes toward intimate artificial companions and social robots in such forms as sex robots and care robots. The question then becomes whether, if we need not and indeed cannot work because robots have replaced us in the labor market, we also want them to replace us in intimate relationships or instead will continue to define achievement in terms of intimate and communal relationships with other humans. From this perspective, the question is whether the end of work will help human relationships to flourish or whether the automation of work is simply a prelude to the automation of intimacy and community.

The answers to such questions may seem obvious—most of us would probably argue that, in the world without work, we would be happy to enjoy more and better intimate and community relationships—but, more careful consideration makes clear that this is far from being the case. According to some sociologists, by far most individuals in Western countries tend to consider intimacy—love, family, and so on—the most important aspect of human life. Thus, for instance, Urlich Beck and Elisabeth Beck-Gersheim (1995) claim that, with the marginalization of traditional religion, love has become a new secular religion in terms of endowing the lives of most people with meaning. Arlie Russell Hochschild (2003, p. 143) points out, however, that intimacy holds this importance only at the level of declaration, while in practice few people devote the majority of their time to intimate relationships. Thus we may claim that work is only a means and that family is the end, but our actions invert these priorities.

There are at least two explanations for this paradoxical situation. One is that we really do consider intimate relationships as precious, but that, owing to the pressures of the labor market, are unable to participate in and enjoy them to the extent that we would like to. The alternative explanation is that we are deceiving ourselves and others when we declare intimacy to be the most important part of our lives. It is difficult to say which of these explanations more accurately accounts for the paradox, making it in turn difficult to predict the status of intimacy in the future or even to answer the question recently posed by Eva Illouz (2017) regarding whether love remains a part of the good life.

It is also worth mentioning in this context that, even if the first option is correct and we would really like to have more time to participate in and enjoy intimate relationships, this does not mean that we must necessarily oppose the automation of intimacy. That is, we may wish to participate in intimate relationships but to do so not with people or not with people only. These considerations recall the examination of attitudes towards sex robots and care robots in the second chapter of this book. Again, I do not offer an opinion as to whether anthropocentric robot enthusiasts' claims that robots will be better than people are more likely to be correct than the claims of anthropocentric enthusiasts and anti-anthropocentric enthusiasts that robots will be better than people but will, rather than replace them, simply join them in the intimate relationships as a supplement rather than substitute of humans. Nor do I know whether anthropocentric skeptics are right in arguing that any intimate relationship between robots and humans will lead to the degradation and disintegration of the latter relationships in general and intimate relationships in particular.

Confronting these issues involves deciding whether we consider robots to be tools or recognize them as persons, whether we should continue disenchanting humans and enchanting robots in a world without work or should invert this tendency and re-enchant human beings instead. As I have tried to emphasize in this concluding chapter and throughout this book, there are no easy answers to these questions, but I am convinced that the difficult process of trying to answer them is worthwhile. This process involves detailed consideration of such future scenarios as the ones discussed here and later on in this chapter, as well as such radical and controversial ideas as Danaher's (2017b) claim that artificial offspring—robots recognized as children—have the potential to provide individual human beings with a greater sense of living a meaningful life than the experience of rearing biological children.

Returning, then, to the meaning of life in a world without work, such a world may seem on the one hand to lack of opportunities for achievement and thus for a fulfilling existence. From this perspective, intimate robots are problematic in that, rather than being demanding, they adjust to the needs of individuals and therefore fail to present the kinds of challenges that enable achievements and create meaning. The kinds of doubts, paradoxes, and arguments discussed in this book—such as Danaher's argument about artificial offspring—however, indicate that matters are not so obvious or simple. I accordingly urge further discussion of issues relating to the status of the need for intimacy robots, both now and in the future, and the extension of this issue to artificial offspring. It is particularly important is to investigate whether consideration of such scenarios has the potential to change the attitudes of the various participants in the discussion, since they might be helpful in clarifying, establishing or changing the already taken positions.

Special attention also needs to be given to community relationships in the context of a life characterized neither by work and nor by economic difficulties. Thus we might ask whether a world without work would mean an end to widespread individualization (Putnam 2000), "culture of narcissism" (Lasch 1979), and the "the fall of public man" (Sennett 1977). Would we like to reunite with our neighbors and enjoy their company or even participate actively in civic affairs? Or would we prefer to avoid our neighbors and the public sphere? Or should we make our public affairs automated as well and look to artificial intelligence to provide, not just a means to our ends, but the ends themselves? Again, answers to these questions are far from obvious.

5.2 Taking Robot Rights Seriously: When Designing Becomes Eugenics

My second example of a far-fetched future scenario concerns the potential unintended consequences of endowing robots with rights. This issue, which has thus far received only brief mention in this book, was recently and thoroughly examined by the aforementioned David Gunkel (2018). If we take it seriously, we should begin taking into account the welfare of both humans and robots, in like manner as the issue of the welfare of animals was raised after we granted them rights. The consideration of the welfare of robots would then be an important part of their development, alongside consideration of human needs. In fact, in this discussion, robot ethics meets bioethics, in that designing robots can be seen as a kind of positive liberal eugenics performed at the level of genetic code. Only a few philosophers have touched on this issue, namely Steve Petersen (2007, 2012), Mark Walker (2006, 2014), and myself (Musiał 2017).

The main issue discussed by Petersen and Walker is whether robots that would be recognized as persons (for whatever reasons) should be designed as servants. While Walker considers such an approach to robot design immoral as a kind of slavery, Petersen asserts that, if thus designed, these robots would be happy serving humans because doing so would be a fulfillment of their desires, which is to say of the desires that were designed for them. This debate has, to my way of thinking, been superficial; for I consider the very act of designing a person problematic, whatever the purpose for which that person is designed, whether to be a servant or a leader. The problem of designing a person is related to the problem of positive liberal eugenics at the level of manipulating genes as discussed in the field of bioethics. In other words, if we decide that robots are persons by endowing them with rights or in any other way, then most of the moral dilemmas connected with positive liberal eugenics become significant for the design of robots as well. It is in this respect that I draw attention, following Jürgen Habermas (2003), to some significant harms that could result from the very fact that a person—whether biological or electronic—has been designed by someone. Once more, I do not consider Habermas's argument or my account of it as conclusive solutions to the problem—my point is rather to suggest some of the unintended consequences that may result from recognizing robots as persons.

Before discussing Habermas's argument in detail, it will be useful to say a few words about the process of design itself. Petersen (2012) discusses design through a thought experiment in which he envisions a machine—the "Person-o-Matic"—that makes it possible to add a particular feature to a designed person with the push of a button. In this experiment, design is conceived of as an intentional process in which one consciously and deliberately selects some features of the designed person, though Peterson attempts to argue that such a process is analogous to "natural" breeding on the one hand and to being raised by parents on the other (pp. 286, 288). Following Habermas, I will distinguish designing from both of the abovementioned phenomena.

I understand "natural" breeding as each case in which a person is born without any intentional choice in reference to his or her features. Natural breeding thus includes regular sexual intercourse, in vitro fertilization, or the use of surrogate mothers; in each of these cases, there are neither any buttons to push nor any individual who might push them if they were available. This is exactly what differentiates intentional design from natural breeding: the lack of intentional choice. Of course, natural breeding is not entirely random and undetermined, since biological laws are in play, though it is difficult to claim that these laws are a result of someone's intention. Moreover, human beings may intentionally seek partners with specific genetic features in hopes of transferring these features to their offspring, though there is no certainty that their efforts will be rewarded. In the case of the Person-o-Matic, on the other hand, the choice of features is guaranteed. In Habermas's terms, then, "natural" breeding happens "by chance," while intentional design takes place "by choice." The notions of chance and the choice here refer, of course, to the selection of features of the person, not to the decision to bring a person into the world. Habermas further asserts that natural breeding involves an element of contingency, while intentional designing does not. These concepts will be discussed in further detail presently, but it is first necessary to examine the distinction between designing children and raising them.

It is hard to disagree with Habermas that intentionality is involved in both designing and raising children. However, the difference between designing and raising is present somewhere else. To show this, Habermas refers to his theory of communicative action and claims that while raising children is based on communicative action, in the case of designing it is replaced by instrumental action. As he puts it, designed features, as "one-sided and unchallengeable expectations," are "genetically fixed 'demands' [that] cannot, strictly speaking, be responded to" (2003, p. 51). On the other hand, raising children is not one-sided but rather mutual, and thus can "be responded to," if not immediately. This mutuality is a consequence of the fact that such communicative actions as raising children are intended to achieve a consensus among the involved parties; the aim of instrumental action, by contrast, is the achievement of one party's goals irrespective of the goals of others. Raising children, then, is about talking to them in the inter-subjective space of dialogue in which they may respond—sooner or later, and more or less efficiently—so that the communicative action in which raising is embedded involves treating the other person as a subject. Designing a person, however, involves "talking to them" in manner in which they will never be able to respond and in which, moreover, there is no interest in any recognition of the other person as a subject; that other person is, rather, an object to be adjusted to the will of its designer.

Therefore, intentional designing differs from "natural" breeding in respect to the intentionality and contingency, and from raising children due to the type of the action involved. According to Habermas, these differences between natural and designed persons are responsible for harms suffered by the latter in terms of self-identity, autonomy, and equality. In his investigations of self-identity, Habermas considers both the mental and physical aspects apropos of beliefs and desires and about the body. He argues that, when we obtain these beliefs and desires "by chance" or as a result of raising and socialization based on communicative action, we have no problem considering them our own precisely because we have received them by accident or agreed on them in the process of communication. If, however, someone designs our beliefs and desires into us, if we feel that they have been given to us intentionally but without our agreement, how can we consider them our own? How can we consider them a part of our identity? Habermas extends this argumentation to the body: if we have received our bodies contingently and "by chance" in the process of natural breeding, we feel that we *are* our bodies; when, however, someone has intentionally designed our bodies, we feel that we only *have* them by the choice of someone else. We cannot, Habermas claims, identify with our bodies when someone has intentionally designed them for us, and this is the harm that designed persons suffer from intentional design with regard to their self-identity: they cannot identify with either their beliefs and desires or with their bodies.

Another argument provided by Habermas against positive liberal eugenics (again, understood as designing the person before he or she is born by genetic manipulation) refers to the autonomy of a designed person and is strictly related to the one that refers to self-identity. If we define autonomy as the ability to follow one's own beliefs and desires, then designed persons may consider their autonomy problematic, since their beliefs and desires are not, or may not be, their own, however paradoxical this may sound. As alluded to above, it can be argued that most of our beliefs and desires are not our own, but the products of genetic heritage and socialization. It is in this context that Habermas points to differences between both intentional design and natural breeding on the one hand and raising children on the other, claiming that contingent beliefs and desires received by chance are not our own, nor are those communicated to us and with which we may on some level disagree. If, however, someone designs these beliefs and desires into us on a genetic level, this is a matter neither of chance nor of communication and agreement. Since we cannot consider these beliefs and desires our own, we cannot consider ourselves autonomous, because we do not follow our beliefs and desires, but someone else's.

Joel Anderson (2005) provides an example of an individual who is naturally—that is, not owing to someone else's intentional design—very tall and is expected to play basketball. Anderson imagines how much more fraught such expectations would be for an individual who has been intentionally designed to be tall. For while the "naturally" tall person might consider being tall a matter of a bad luck and reject expectations regarding playing basketball, the "designed" tall person would aware that being tall was the result of someone else's choice and thus that expectations based on physical appearance were a matter of biological heritage and could not be so blithely rejected. In Anderson's example, then, the extent to which the tall person's autonomy is limited depends on whether tallness resulted from a conscious choice or design.

A last claim of Habermas relevant to the present discussion is that designed persons would suffer from inequality, specifically that the domination of parents over children could not become inverted. Traditionally, parents have at first taught children about the world and later been taught by them about it—for example, younger people may teach older ones how to use technological devices; so also, parents provide financial support and care to young children and then later receive financial support and care from their adult children. Habermas asserts, however, that there is no adequate response to their parents for children who have been intentionally designed, since such parents will have performed instrumental rather than communicative action; the children's communicative actions would not be equal with respect to the manner in which the parents "started the conversation" by designing them. Under these circumstances, the parents would always remain intentional designers of the children and their features, while children will always remain their product that cannot effectively respond to the fact of being designed in a particular way. Obviously, such inequality is not something entirely new, since the fact that parents bring their children into life and children cannot do the same think to their parents is unavoidable. However, Habermas points out that intentional designing adds another level to this inequality, and that this level—in contrast to the basic and unavoidable one that is mentioned above—is not necessary.

It is worth mentioning, that there is also another level of possible inequality, or maybe even exclusion, that Habermas does not investigate. It refers to the differences between those who were designed and those who were not designed by their parents. On the one hand, this latter group can consider the member of the former as bizarre, unnatural, incomplete human beings; on the other hand due to the fact that designing will implement positive features, those who will be designed can consider the rest as worse, and may even try to compensate the issues with self-identity and autonomy by stigmatizing those, who were not designed. Choosing the correct scenario is far from obvious, however it is important to consider the economic factor: designing will probably be expensive and due to that fact only the richest people will be able to afford it. So, being designed may be another way in which the rich will try to establish their domination over the poor, or a reason for which the poor will stigmatize the rich. Therefore, the fact of being designed may result not only in the increase of inequalities between parents (designers) and children (product), as Habermas emphasizes, but may also entail various fluctuations of social inequalities between rich and designed on the one hand, and poor and not designed on the other.

If, then, we take into account some of the arguments against positive liberal eugenics, we may be dissuaded from designing robots intentionally at all on the grounds that doing so would harm their personhood and deny their rights. The only alternative for producing robots would then be to do so in a random way; but such robots would cease to enchant us, since part of their magic is that we can develop them intentionally and adjust them to our needs. If robots were to become persons and be granted rights, this form of intentionality might be considered unethical. Therefore, the consequence of ascribing personhood and rights to robots—that is, of enchanting them in radical ways—might be that the robots would become disenchanted, just like contemporary human beings are becoming disenchanted.

5.3   Final Remarks

The two scenarios just discussed (and investigated in more details by Danaher, Gunkel, Petersen and myself elsewhere) are useful for posing questions about the future that need to be taken seriously, not only, or even primarily, for their predictive power. Rather, these scenarios are useful because they help to shed light on the present, on the ideas that we have, the actions that we perform, and the tendencies that we can observe. Just as when we wonder about the future of education, we also start to wonder about its contemporary function, so also, when we think about artificial offspring, we start to think about natural children. In sum, when we ask about who we might become and what will give our life meaning, we also start to ask who we are right now and how we make can our rights meaningful at this moment.

Considering these scenarios from the perspective established in the previous chapters of this book, we may ask whether endowing robots with rights and enabling widespread automation would more likely entail increased re-enchantment or instead progressive disenchantment. More specifically, we may ask whether, in a world of widespread automation and without work, we should continue disenchanting humans and enchanting robots, or rather, perhaps, we should re-enchant humans. Also useful would be a discussion of whether endowing robots with rights would mean making them and the experience of interacting with them more magical or, on the contrary, would mean rationalizing and disenchanting everything that is magical about them.

One might of course argue to the contrary that discussion of these scenarios is counterproductive given the copious problems—data mining, loss of privacy, and so on—relating to contemporary technology and especially AI. I fully acknowledge that such problems are far more pressing than abstract issues examined in this book, and I would not wish to distract anyone from directing attention to these problems. Nevertheless, it is my conclusion that one of the sources of our current problems is a tendency to develop and market new technologies without first considering how to regulate them in light of obvious issues as well as potential unexpected problems and unintended consequences. It might be better to think first—even if prematurely—and then to decide whether we truly want something rather than to decide without thinking and then be beset with worries. My hopes in this regard are, though, likely naive.

To summarize this inconclusive conclusion, it is my contention that, in order to understand where we are now and what lies before us, we ought to consider, not only where we came from and how much of what we think we have left behind may still be with us, but also where will we go from here. It is my belief that, in doing so, we will take into account not only our ideas, actions, and intentions but also structural and non-human factors that determine what we think and what we do and the unintended consequences of our ideas and actions. We should consider, therefore, both where we want the future to take us and futures that we wish to avoid as well as where the future may take us despite our wishes.

References

Anderson, J. (2005). Habermas, Jürgen. The Future of Human Nature (Book Review). *Ethics, 115*(4), 816–822.

Beck, U., & Beck-Gernsheim, E. (1995). *The Normal Chaos of Love.* (M. Ritter & J. Wiebel, Trans.). Cambridge, UK: Polity Press.

Danaher, J. (2017a). Building a Post-Work Utopia: Technological Unemployment, Life Extension, and the Future of Human Flourishing. In K. LaGrandeur & J. J. Hughes (Eds.), *Surviving the Machine Age: Intelligent Technology and the Transformation of Human Work* (pp. 63–82). London: Palgrave Macmillan. https://doi.org/10.1007/978-3-319-51165-8_5.

Danaher, J. (2017b). Why We Should Create Artificial Offspring: Meaning and the Collective Afterlife. *Science and Engineering Ethics,* 1–22. https://doi.org/10.1007/s11948-017-9932-0.

Danaher, J. (2017c). Will Life Be Worth Living in a World Without Work? Technological Unemployment and the Meaning of Life. *Science and Engineering Ethics, 23*(1), 41–64. https://doi.org/10.1007/s11948-016-9770-5.

Gunkel, D. J. (2017). Rage Against the Machine: Rethinking Education in the Face of Technological Unemployment. In K. LaGrandeur & J. J. Hughes (Eds.), *Surviving the Machine Age: Intelligent Technology and the Transformation of Human Work* (pp. 147–162). London: Palgrave Macmillan. https://doi.org/10.1007/978-3-319-51165-8_10.

Gunkel, D. J. (2018). *Robot Rights*. Cambridge, MA and London: MIT Press.

Habermas, J. (2003). *The Future of Human Nature* (H. Beister, M. Pensky, & W. Wehg, Trans.), Cambridge, UK: Polity Press.

Hochschild, A. R. (2003). *The Commercialization of Intimate Life: Notes from Home and Work*. Berkeley: University of California Press.

Illouz, E. (2017). Is Love Still a Part of the Good Life? In H. Rosa & C. Henning (Eds.), *The Good Life Beyond Growth: New Perspectives*. Abington and New York: Routledge.

Lasch, C. (1979). *The Culture of Narcissism: American Life in an Age of Diminishing Expectations*. New York and London: W. W. Norton.

Musiał, M. (2017). Designing People to Serve—The Other Side of the Coin. *Journal of Experimental & Theoretical Artificial Intelligence, 29*(5), 1087–1097. https://doi.org/10.1080/0952813X.2017.1309691.

Musiał, M. (2018). Automation and the Meaning of Life: A Sense of Achievement and Being with Others. In M. Coeckelbergh, J. Loh, M. Funk, J. Seibt, & M. Nørskov (Eds.), *Envisioning Robots in Society—Politics, Power, and Public Space: Proceedings to Robophilosophy 2018 / TRANSOR 2018* (pp. 251–258). Amsterdam, Berlin, Tokyo, and Washington: IOS Press.

Petersen, S. (2007). The Ethics of Robot Servitude. *Journal of Experimental & Theoretical Artificial Intelligence, 19*(1), 43–54. https://doi.org/10.1080/09528130601116139.

Petersen, S. (2012). Designing People to Serve. In P. Lin, G. Bekey, & K. Abney (Eds.), *Robot Ethics: The Ethical and Social Implications of Robotics* (pp. 283–298). Cambridge, MA and London: MIT Press.

Putnam, R. D. (2000). *Bowling Alone: The Collapse and Revival of American Community*. New York: Touchstone Books by Simon & Schuster.

Sennett, R. (1977). *The Fall of Public Man*. New York: Knopf.

Walker, M. (2006). A Moral Paradox in the Creation of Artificial Intelligence: Mary Poppins 3000s of the World Unite. In *Human Implications of Human-Robot Interaction: Papers from the AAAI Workshop*, 23–28. http://www.aaai.org/Papers/Workshops/2006/WS-06-09/WS06-09-005.pdf. Accessed 16 October 2018.

Walker, M. (2014). BIG and Technological Unemployment: Chicken Little Versus the Economists. *Journal of Evolution & Technology, 24*(1), 5–25.

INDEX

The manufacturer's authorised representative in the EU is Springer
Nature Customer Service Centre GmbH, Europaplatz 3, 69115 Heidelberg,
Germany. If you have any concerns regarding our products, please
contact ProductSafety@springernature.com

Printed and bound by CPI Group (UK) Ltd, Croydon, CR0 4YY
29/04/2026
02099478-0018